THE CHEMICAL SOCIETY
MONOGRAPHS FOR TEACHERS No. 31

The Principles of Bio-inorganic Chemistry

ANNA M. FIABANE, BSc, PhD
Ninewells Hospital, Dundee

and

DAVID R. WILLIAMS, BSc, PhD, CChem, FRIC, DSc
University of Wales Institute of Science and Technology, Cardiff

LONDON: THE CHEMICAL SOCIETY

Monographs for Teachers

This is another publication in the series of Monographs for Teachers which was launched in 1959 by the Royal Institute of Chemistry. The initial aim of the series was to present concise and authoritative accounts of selected well-defined topics in chemistry for those who teach the subject at GCE Advanced level and above. This scope has now been widened to cover accounts of newer areas of chemistry or of interdisciplinary fields that make use of chemistry. Though intended primarily for teachers of chemistry, the monographs will doubtless be of value also to a wider readership, including students in further and higher education.

First published August 1977

Published by The Chemical Society, Burlington House, London W1V 0BN, and distributed by The Chemical Society, Distribution Centre, Blackhorse Road, Letchworth, Herts SG6 1HN

Printed by Adlard & Son Ltd, Bartholomew Press, Dorking

CONTENTS

1. Introduction

The laws of nature are imperative and irrevocable; one such law is that the development of any part of the human body is entirely dependent upon the nutritional conditions in which the component grows. Although it has been known for many years that adequate quantities of many organic constituents (for example, sugars, proteins and vitamins) and perhaps one or two inorganic ingredients (especially iron and calcium) were necessary for health, our requirements for many other elements were overlooked because they were present in man in such low concentrations. Fortunately, soil impurities clinging to food generally provided a reasonable spectrum of such elements but deficiencies sometimes occurred.

More recently, metal deficiency diseases have become more widely known and this has focussed the attentions of laboratory researchers in biochemistry, inorganic chemistry, medicine and pharmacology upon the exact determination of elemental concentrations (in order to establish the elemental status of a patient) and on the mechanisms involved in the bio-availability, assimilation and excretion of trace elements. Even more recently pathological conditions arising from trace element excesses have come to light, frequently because a polluted environment causes symptoms in people living or working in a specific area. Often these conditions are not mirrored in wild animals because, living in a natural habitat, they tend to acquire their trace elements from impurities on their food and so avoid the deficiencies or excesses of our standardised 'purified' foods which can lead us to the so-called 'civilisation diseases'.

This Monograph is not advocating our return to precivilisation diets (*a*) because the soil content in any given area, as reflected in dietary intakes, is usually *sufficient* in *most* elements but *deficient* in *one* or *two* others, (*b*) because natural, unprocessed foods, sometimes contain challenges to life in the form of bacteria or viruses, and (*c*) because modern soils contain many impurities that can be dangerous (herbicides from crop spraying, heavy metals concentrated into sewage sludge *etc*). Thus, it appears desirable to have a combination of all the 'best' traces of soil elements taken from many areas. This is judged from a complete element analysis of human requirements, from the bio-availability of these elements from palatable diets, and from the degrees to which these ions require supplementing. This brings into court the whole question of normal, or reference, values and embodies such familiar principles as labile and inert metal complexes, oxidation states, the roles of these elements *in vivo*, the symptoms and treatment of bio-inorganic diseases, and

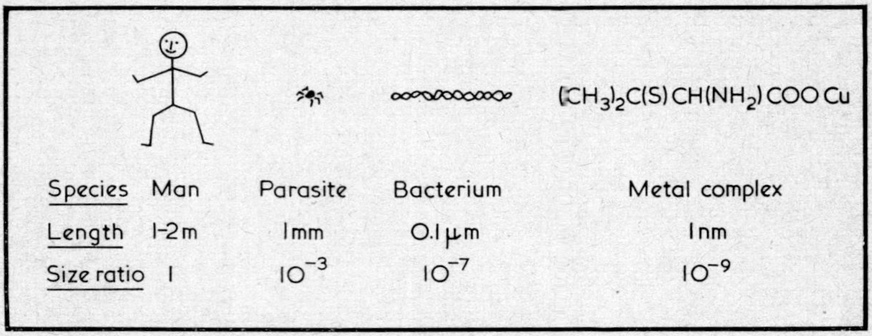
(CH3)2C(S)CH(NH2)COO Cu
Species Man Parasite Bacterium Metal complex
Length 1–2 m 1 mm 0.1 μm 1 nm
Size ratio 1 10−3 10−7 10−9

FIG. 1. The health of *homo sapiens* is under the influence of ligands and metal complexes weighing 5×10^{-27} times that of the human body.

so on, and it is the aim of this Monograph to inform the reader of the current state of the art.

Thus, we can see that a general definition of bio-inorganic chemistry is 'a branch of physical science whose aims are to understand the chemistry of reactions involving the essential metals, and other trace elements, *in vivo* and to apply this knowledge'.

Figure 1 shows that to consider the influence of metal ions, ligands and complexes upon our health we must be able to envisage considerable changes in scale in terms of the relative sizes and masses of species. This arises because humans are dependent upon the existence of species which are far smaller than themselves and these, in turn, are dependent upon even tinier species.

We shall see in the next chapter that there are two broad groups of elements in the human body—those involved in health (the essential and beneficial ones on p 7)—and those involved in interrupting healthy reactions (the contaminating and polluting elements). Indeed, the majority of chemical elements can be found in minute quantities in the human body, their concentrations reflecting the concentrations in the food, soil and atmosphere. Each of these elements exhibits a spectrum of biological responses which is dependent upon the concentration of the element or its compound in any particular organ or body fluid. *Figure 2* shows how the state of health of an organ is dependent upon the concentration of an element in that organ.

As the concentration of a useful element is increased from A to B the biochemical response becomes progressively more normal until plateau B to C is reached. This 'concentration for optimum health' plateau varies from essential/beneficial element to element and from organ to organ and its width is determined by the homeostatic capacity of the animal or system. Region C to D depicts the decline

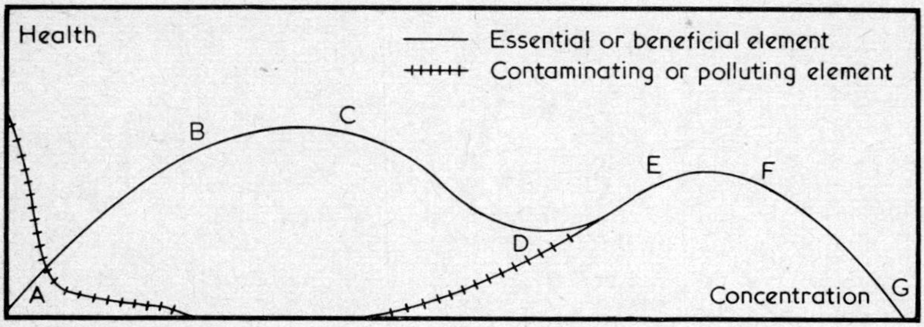

FIG. 2. Healthy biological response plotted *versus* the concentration of any element *in vivo*.

in health as excessive concentrations build up in an organ (for example, the siderosis effects associated with too much iron) until the curve reaches trough D.

Some diseases, especially those caused by invading organisms such as viruses or bacteria, do not respond adequately merely to correcting element *in vivo* concentrations to plateau B to C. In these instances there may be grounds for administering a higher concentration of a metal or of its complex. Region D to E shows the pharmacological effect of administering this element as doses of drugs. Such drugs stimulate the host's defence mechanisms. Naturally there is a limit to this process at plateau E to F. We are fortunate that this plateau exists because it means an 80 kg man can be reasonably safely prescribed the same quantity of drug as a 50 kg girl. All therapies eventually change from excitation to toxicological inhibition of the living process. F to G depicts this drug poisoning. Eventually, large doses of the element—as an essential element, as a polluting element, or as a medically administered compound of the element—cause an irreversible reaction, a complete decline in the living systems, leading to death.

Such curves differ from element to element, some having better homeostatic capacities than others and some having but a brief safety margin between optimum and toxic concentrations. Further, such *curves* ought really to be *surfaces* enclosed in three dimensional axes since healthy concentrations of elements are sometimes dependent upon the prevailing concentration of other elements. (For example, there are widely known mutual antagonisms between Fe and Co, Cu and Mn, Cu and Mo, Cu and Zn, and Ca and K concentrations.) Finally, we must remember that these curves or surfaces may have a varying amplitude according to prevailing circadian rhythms (*ie* ionic concentrations vary on a 24 h basis).

Samples of animal tissue collected from different corners of the world can be found to have more or less a constant elemental composition. This is a puzzling natural phenomenon because, whereas plants and some animals can synthesise many of the vitamins whenever required, the element must be acquired from an environment (air, soil and water) in which they are by no means evenly distributed; and so, whereas biochemistry texts contain many references to biosynthetic pathways to vitamins, hormones, *etc*, bio-inorganic mechanisms usually commence with considerations of bodily intake and questions concerning bio-availability. Another difference between 'organic' biochemistry and bio-inorganic chemistry is that the bio-inorganic elements have their maximum influence in very early life at which time the pattern of the future development is established. Thus, foetuses and plant seeds have richer concentrations of trace elements than found in mature animals and plants.

On the one hand, man is fortunate in that he is at the end of the food chain and so the plants and animals coming before man have accumulated many of the elements that he needs. On the other hand, when plants or animals accumulate toxic elements (for example, arsenic, cadmium, lead or mercury) these become concentrated as they pass down the food chain and can seriously detract from man's health: this leads to a varied pattern in the distribution of diseases.

Some relationships between the geochemical environment and health or disease have been well known for decades, for example the relationship between iodine and thyroid dysfunctions. However, more recently, we have come to realise that this is just the tip of the iceberg, there being many other conditions which are also trace element dependent.

Disease, of course, has many causes—genetic, viral, bacterial, geochemical and so on—but the beauty of the geochemical influences lies in the fact that they remain *in situ* long after the disease pattern has been discovered and the epidemic has passed. Furthermore, new research is continually expanding the number of metabolic symptoms known to be influenced by trace elements and is revealing many hitherto unsuspected relationships between element concentrations and abnormal states of health. Thus, soon the evaluation of such concentrations in various organs will be used as a means of diagnosing illness. So too the manipulation of these concentrations will play an even greater role in its prevention. This prophylactic manipulation may either involve spreading inorganic 'fertilisers' on the soils or administering ligands and complexes as drugs* chosen to correct the element imbalances.

---

* The word 'drug' has recently taken on a more sinister connotation; nowadays it implies 'dangerous chemical' or 'addictive agent'. However, we use the word in its true sense, *ie* a chemical used in the prevention and in the treatment of disease.

We hope that this Monograph will throw into sharp relief the intriguing relationships between the production of inorganic drugs for treating disease and the inorganic imbalances in nature that underlie such diseases, and that chemists will become acquainted with areas of current interest in biology and medicine. Since many readers will be chemists, Chapter 3 reviews the biology of some of the more important bio-inorganic elements. Life scientists may prefer to omit this chapter. Attention is focused on the principles of the subject rather than on soporific lists of details but the references have been selected to lead the reader to these facts when necessary. Anyone requiring a revision of the biochemistry of man is referred to a sister volume in this series (E. G. Brown, *An introduction to biochemistry*, Monograph No. 17).

2. The Origin of the Elements comprising Biological Systems

The elements of life

In 1828 Friedrich Wöhler exploded the myth that organic molecules could be produced only in living systems when he synthesised a biological molecule, urea, from an inorganic source. We shall, in the main, restrict our discussions to the inorganic chemistry of the elements occurring in the human species but, nevertheless, we shall readily refer to animals or plants if superior examples for illustrating a bio-inorganic principle are more evident in those species.

The Periodic Table of some 90+ stable elements sets an absolute limit to the number that may be found in an organism. Even after more than a century of research, the exact number that are necessary for healthy life is unknown. As increasingly refined micro-analytical techniques and procedures for obtaining element-sterile diets and environments are developed, this list of bio-inorganic elements increases. Thus, the classified lists of elements to be discussed in subsequent paragraphs ought to be regarded as a still picture of what is, in reality, an exciting, developing movie film. From past experience we expect that the principles of bio-inorganic chemistry will apply to the essential or beneficial elements yet to be discovered.

There are three characteristics of the biosphere which have played a major role in deciding the chemistry of living species:

(*i*). The ubiquity of water on the earth's surface and the peculiar chemistry that arises from the unexpectedly high stability and melting point of the H_2O molecule: this has divided evolution into those reactions that occur in the solution phase and those that occur catalysed at the surfaces of precipitates.

(*ii*). Although silicon is 146 times more plentiful than carbon on the earth's surface, nature has chosen carbon because of the stability and relative water solubility of carbon dioxide and because of the ability of carbon to form –C–C–C– rings and chains (C–C bonds are more stable than Si–Si bonds).

(*iii*). Apart from the carbon *versus* silicon-type anomalies just mentioned, the ionogram (a graph, or plot, of the distributions of ions, or elements) of elements in man reflects the distribution of elements on the surface of the earth, in the oceans and in the lower atmosphere. Because the heavier elements have been compacted into the molten core of the planet, we are composed of the lighter elements of the Periodic Table.

There are four groups of elements in the human body—the *essen-*

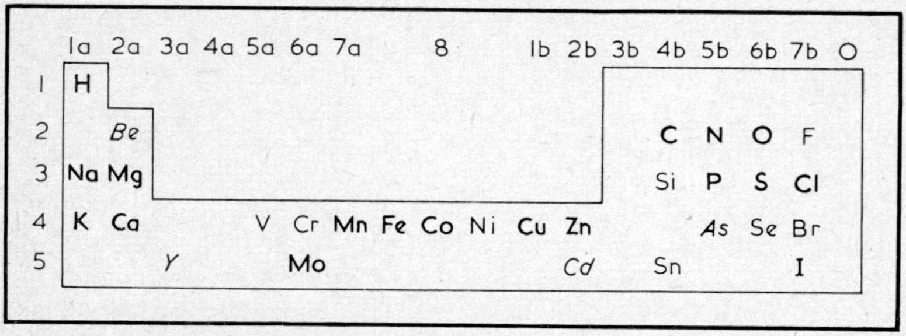

Ia 2a 3a 4a 5a 6a 7a 8 Ib 2b 3b 4b 5b 6b 7b O
1 H
2 Be C N O F
3 Na Mg Si P S Cl
4 K Ca V Cr Mn Fe Co Ni Cu Zn As Se Br
5 Y Mo Cd Sn I

FIG. 3. The **essential**, beneficial and some *carcinogenic* elements for man.

tial, *beneficial*, *contaminating* and *polluting* elements. The adjective 'essential' is only used (*i*) if the element is present in all healthy tissues, (*ii*) if it has a fairly constant concentration range across species, and (*iii*) if its exclusion from the body causes reproducible physiological abnormalities which are reversible upon readmittance of the element. Such rigid definitions have, to date, restricted the essential elements to 18 in number. There are currently also some eight beneficial elements known (*see Fig. 3*). Life is possible without these elements but it is a fairly meagre form of existence that could not be called healthy (*see Fig. 4*). In addition there are 20–30 trace

FIG. 4. The rat on the right of this photograph has been maintained on a fluorine, tin and vanadium free diet for 20 days whereas the rat on the left had a normal diet. (From E. Frieden, *Scient. Am.*, 1972, **227**, 52, by kind permission.)

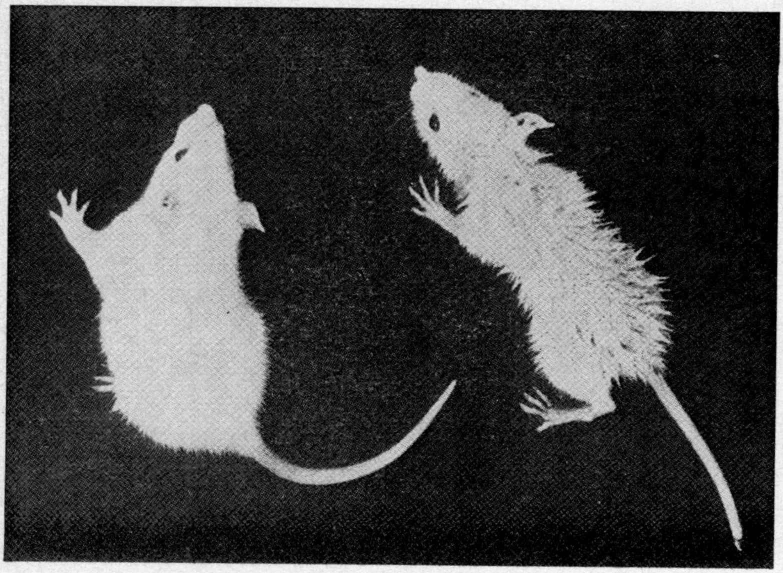

elements which are ubiquitous in tissues but whose concentrations vary and whose physiological roles have not yet been fully determined. These must be called contaminants until shown to be otherwise. If the concentration of a contaminant resembles that at which physiological or behavioural symptoms can be detected then the contaminant may be called a pollutant (for example, very low concentrations of lead, cadmium or mercury in blood can have deleterious effects and so can be called pollutants). The dividing lines between essential and beneficial and between contaminating and polluting elements are not definitely fixed and are expected to be redrawn from time to time as diagnoses and instrumentation improve.

Each element in each of the four groups *in vivo* has a cycle in nature. (For example *see Fig. 5.*) These interrelationships are similar to those found in the familiar carbon or nitrogen cycles. The initial availabilities of the elements and the relative kinetics of the processes embodied in each stage of these cycles have dictated that species have constant compositions (*see* Table 1) in each organ.

Table 1. Average element composition of a 70 kg human.

Element	g man^{-1}
Main group metals	
Sodium	70
Potassium	250
Magnesium	42
Calcium	1700
Transition series metals	
Manganese	<1
Iron	6
Cobalt	<1
Copper	<1
Zinc	1–2
Molybdenum	<1
Nickel	<1
Main group non-metals	
Hydrogen	6580
Carbon	12 590
Nitrogen	1815
Oxygen	43 550
Phosphorus	680
Sulphur	100
Chlorine	115
Iodine	<1

There are two supplementary points to be mentioned in this connection: the situation as we observe it in today's snapshot is not static. Races of people probably will exist in, for example, countries which are rich in transition element ores, or in future generations, who will have concentrations of some elements different to those listed in Table 1. However, it takes centuries to achieve such

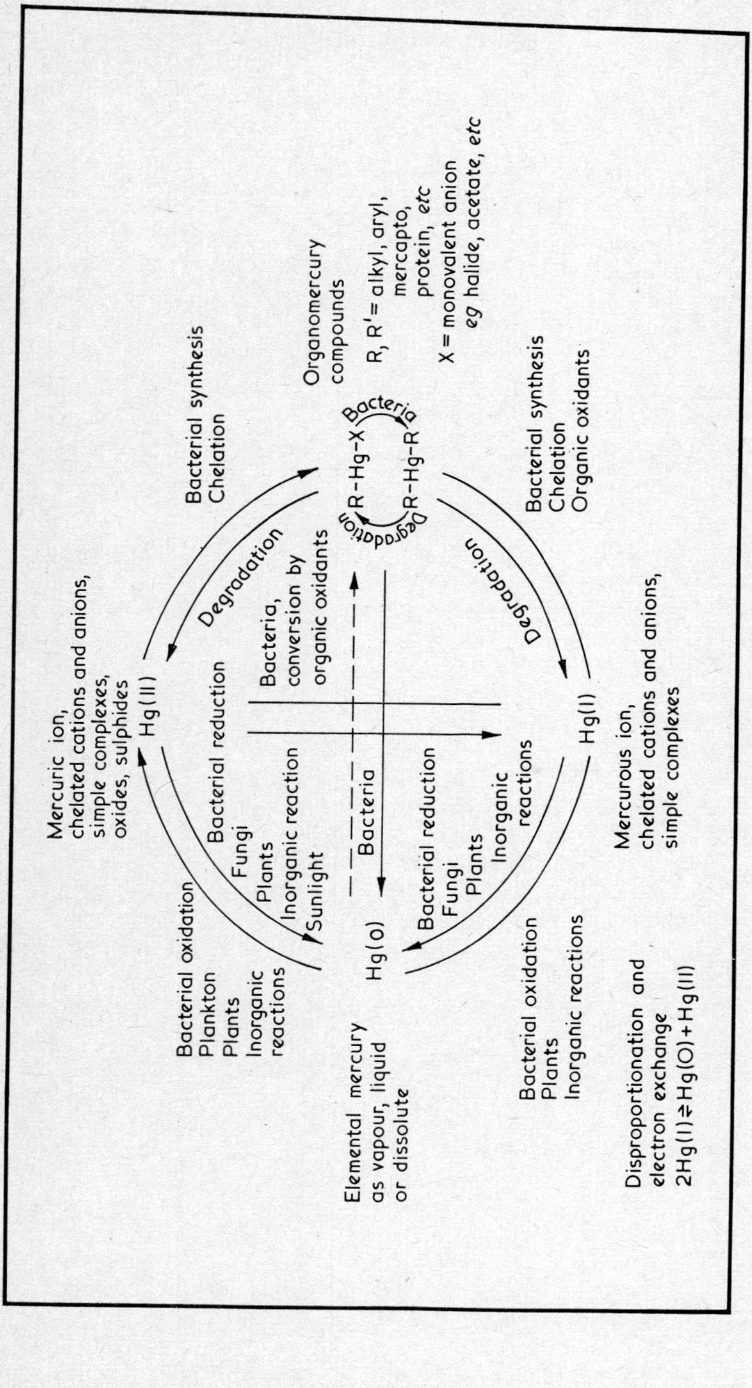

FIG. 5. The mercury cycle.

changes. The second point is the converse of this observation: if we substantially increase the concentration of an element in our diet it will probably have an adverse effect upon our health. Fortunately, nature has equipped us with means of excluding or excreting *moderate* excesses of the essential elements, but less so for the non-essential elements—thus poisoning from these latter is more likely. Finally, we might note that if a polluting or contaminating element enters our body its biological half-life may be long because we do not have a pathway for its efficient excretion.

Evolutionary aspects of metal ions *in vivo*

Most scientists are in agreement that our planet was formed as the condensation product from gas and dust particles from some immense supernova explosion in outer space some 4.5–4.7 × 10^9 years ago. The whole question of the evolution of life is a contentious one. This controversy arises, in the main, because of a lack of scientific data on how early cells developed: microbial evolution is the furthest back we can go. (They are at the plateau's edge from which one has to use pure philosophy to move back in time.) This new planet had a hard core and a reducing atmosphere of H_2O, H_2S, NH_3, CH_4 and perhaps some CO_2. Over the next 10^9 years these molecules were bombarded with energy, for example from the sun or from nuclear changes and so reacted to form simple organic species. These, in turn, reacted with inorganic compounds to give monomeric biochemicals (amino acids, nucleotides and sugars *etc*), then biopolymers and so eventually to the first primitive cell which needed a further 3.5 × 10^9 years to evolve into life as we know it today (*see Fig. 6*). This protracted evolutionary process was a continuous progression from primitive, inefficient, mechanisms to more complex, efficient ones. The reactions all occurred in the oceans (or on their shorelines) and metal ions must have played a key role both in determining the compositions of the biopolymers we know today and in dictating whether L or D configurations of sugars and amino acids would be preferred. The principle of tides producing anhydrous conditions on beaches dried by sun and wind and of the reactions occurring in the structural grooves of beach crystals such as silicates or apatites (calcium phosphates) seems a logical suggestion. When monomers condense into polymers dehydration is necessary and this is extremely unlikely in the oceans themselves unless a heterogeneous catalyst is present.

This first cell had about 100 different protein molecules (compare this with modern cells containing many millions of proteins) and it also contained a variety of metal ions, some fulfilling a structural or osmotic role, others acting as catalysts. Magnesium would have been particularly good in this latter role since it is known to catalyse

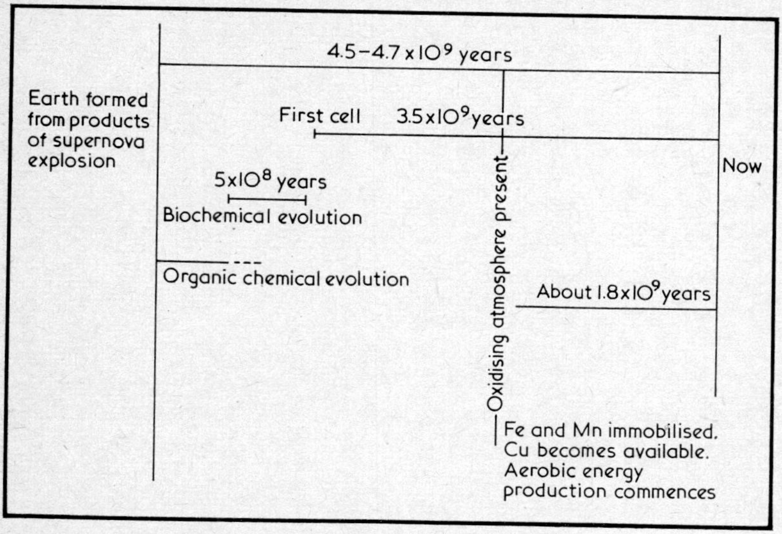

FIG. 6. The evolution of modern man.

condensation reactions and to have been present in high concentrations in primeval seas (it is currently present at 50 mmol dm^{-3} in our oceans). The fact that the ionic composition of many of our body fluids closely resembles that of sea water strongly supports the suggestion that life evolved in, and from, the oceans.

The elements that are biologically important are, in Periodic Table terms, relatively light (*see Fig. 3*), these being the ones near the surface of the earth. The essential bio-metals have been shown to be Na, Mg, K, Ca, Mo, Mn, Fe, Co, Cu, and Zn—note that their ions all have potentially powerful complexing capabilities and that several have important redoxing properties. Since earth's early atmosphere was a reducing one, Mn(II) and Fe(II) would have been important in the first cell systems. The beneficial elements listed (Si, V, Cr, Ni, Se, Br, Sn and F) are thought to have become involved in more 'recent', more highly organised forms of life.

It is interesting to find that the human body has not yet evolved as effective a series of protective mechanisms to guard against overdoses of the beneficial elements as those functioning against overloading of essential metals. For example, 0.1 ppm of Se is beneficial whereas 10 ppm is carcinogenic. If we recall that the Periodic Table is founded upon the concept of relationships between elements, we might understand that the greatest challenges to the processes occurring in normal cells arise from elements in Periodic Table positions adjacent to the essential and beneficial elements (*eg* see carcinogenic elements in *Fig. 3* or compare pollutants Pb, Cd and Hg with Zn and Sn). It appears that these dangerous metal ions simulate the mechanisms through which *in vivo* trace elements act and hence exclude the essential elements.

There are several principles which have dictated that we are what we are in terms of our metal ion content. The major part has been played by the abundance of metal ions in the hydrosphere but, when a metal ion can redox, it is imperative to have an adequate supply of matched reducing or oxidising ligands to maintain the oxidation state which is a part of a biological mechanism (*eg* porphyrins to complex Fe(II) in the haem moiety). When oxygen evolved into the atmosphere (mainly from photosynthesis by blue–green algae) it was very toxic to most living systems because it tended to change the oxidation states of Mn(II), Fe(II), Co(II) and Cu(I) *etc.* However, evolution soon adjusted the cell to protect it against oxidation (mainly by producing peroxisomes which turn O_2 into H_2O_2 then H_2O). In early days, copper was mainly trapped as insoluble Cu(I) sulphides but once an oxidising atmosphere arose, Cu(II) could be added to the list of primitive essential trace elements comprising Mg, Fe and Mn. At about the same time Fe and Mn became immobilised by the reactions Fe(II) $\rightarrow$ Fe_3O_4(s) $\rightarrow$ FeOOH(s) and Mn(II) $\rightarrow$ Mn_3O_4(s) $\rightarrow$ MnO_2(s) and the atmospheric ozone content crept up to 1 per cent which is sufficient to screen out the harmful effects of the sun's ultraviolet radiation, which is apparently more destructive to aerobic metabolism than to anaerobic. This permitted anaerobic metabolism to be replaced by aerobic metabolism, some species using iron and some the newly released copper(II) for oxygen assimilation.

There were certain evolutionary restraints that arose because of competitive complexing. For example, consider Ca^{2+} and Cu^{2+} ions—when Ca^{2+} became available to the evolving system it was rapidly assimilated into enzymes, being bound by carboxylate groups. Cu^{2+}, on the other hand, complexed far more firmly to the host of amino acids inside cells (Ca^{2+}–amino acid interactions are weak) and thus it was only when proteins had developed Cu(II) binding sites powerful enough to overcome the fierce amino acid/Cu(II) complexing opposition that Cu(II) proteins emerged. The development of these Cu(II) specific sites required many generations of mutation. Even today this free amino acid 'selectivity' principle is at work in complexing polluting and contaminating metal ions and so rendering them less harmful to our system than if they were complexed to a protein site.

The main group metal ions Na^+, K^+, Ca^{2+} and Mg^{2+} have closed shell electronic structures and so their roles *in vivo* have evolved to give the best use of their electrostatic bonding properties rather than their covalence characteristics. Thus, their charges and ionic radii have been important in bio-inorganic evolution. For example, they are strongly aquated as ions and so, (*i*) when they complex many molecules of water are shed and so bonds to a ligand are entropy stabilised, and (*ii*) the highly ordered inner and outer solvation spheres of say Mg^{2+} and its high concentration inside cells helps to produce the highly organised system that cells exhibit.

It is interesting to find that Mg^{2+} ions have been in cells since the beginning whereas Ca^{2+} ions are only important in the later, more sophisticated species (for example, those requiring nerve transmission). Because Ca^{2+} is a 'newer' element (as are the beneficial elements mentioned earlier), albeit essential, man has not completely adapted or evolved to handle it and so its precipitates cause problems (cataract, atherosclerosis, gall stones *etc*) which are all too commonly seen.

Two concluding remarks are apposite—(*i*) the specificity of metal ions in the human body is extremely good (that is why many of us will work smoothly for 70+ years without need for service, oil change or spare parts). (*ii*) Researchers into the origins of life and chemical evolution may have become overfascinated with the organic chemical aspects. From the time when simple molecules washed on to beach crystals and were condensed into stereoisomers, the organic and the inorganic aspects of chemical evolution have progressed hand in hand.

Environmental intrusions

For many years, scientists have realised that trace concentrations of the essential and beneficial elements are needed by the human metabolism whereas other elements such as cadmium, lead or mercury at similar concentrations are highly toxic and yet still others, such as selenium, can be both beneficial and toxic within a fairly narrow concentration range. However, it is only very recently that bio-inorganic chemists have begun to probe the great complexity of the interactions between contaminating elements and essential elements involved in the living process. The time may well be fast approaching when determining all trace element concentrations will play a fundamental part in the diagnosis of illness and, indirectly, in its prevention.

All living species exhibit complicated relationships with each other and with their organic surroundings. We have just described how living organisms have evolved to suit their environments and to protect themselves against naturally occurring concentration changes. However, the sudden changes created by human exploitation of nature and modern technology have greatly overwhelmed such evolutionary protection methods and in some instances have become

2

intolerable to life. For example, in nature rocks and soils are eroded by rivers and so sea organisms are bombarded with elements near river estuaries. Just recently, man has been mining out equivalent quantities of metals and many of these eventually end up in the seas or the atmosphere above them—an extremely heavy burden (*see* Table 2). Several of the elements in the table are essential and yet, as

Table 2. The estimated annual addition of metals to our environment (kg/y).

	Geological dissolution by rivers	Additions from industrial mining
Essential metals		
Copper	3.9×10^8	4.6×10^9
Iron	2.5×10^{10}	3.3×10^{11}
Manganese	4.5×10^8	1.6×10^9
Molybdenum	1.3×10^7	5.8×10^7
Zinc	3.8×10^8	4.0×10^9
Beneficial metal		
Tin	1.5×10^6	1.7×10^8
Contaminant metal		
Silver	5×10^6	7×10^6
Polluting metals		
Lead	1.8×10^8	2.3×10^9
Mercury	3×10^6	7×10^6

may be seen from our discussion of *Fig. 2*, even these elements can be harmful in excess. Of course, *ligand* environmental pollution can also dangerously affect the concentration of *in vivo* essential elements. A current example is the use of the nitrilotriacetate ($N(CH_2CO_2^-)_3$) as a phosphate substitute in American detergents.

The most dangerous heavy metal ions, cadmium, lead and mercury, exert their effects upon man mainly as potent enzyme inhibitors. This occurs because the ions are readily polarisable and so the order of binding strength to protein donor groups is S > N > O. Further, the widespread occurrence of phosphate and chloride ions *in vivo* causes the precipitation of insoluble lead hydroxo phosphate and of only slightly soluble mercuric chloride.

Clearly, prevention is better than cure and currently there is a general community awareness of the dangers of heavy metal poisons which is matched only by the tremendous financial pressure to continue using tetraethyl lead as a petroleum additive or methyl mercury dicyanodiamide as a fungicide in agriculture. Regrettably, cases of poisoning still occur (Table 3). The effects of toxic metals on species are related to the rates of their absorption, distribution, positioned or assimilation into an enzyme, and excretion. Inorganic lead exhibits its main effect as blood disorders caused by its combination

Table 3. Some metal dependent conditions.

Essential or beneficial element	Disease arising from deficiency	Disease associated with an excess of the element
Calcium	Bone deformities, tetany	Cataracts, gall stones, atherosclerosis
Cobalt	Anaemia	Coronary failure, polycythemia
Copper	Anaemia, kinky hair syndrome	S.A.K. Wilson's disease
Chromium	Incorrect glucose metabolism	
Iron	Anaemias	Haemochromatosis, siderosis
Lithium	Manic depression	
Magnesium	Convulsions	Anaesthesia
Manganese	Skeletal deformities, gonadal dysfunctions	Ataxia
Potassium		Addison's disease
Selenium	Necrosis of liver, white muscle disease	Blind staggers in cattle
Sodium	Addison's disease, stoker's cramps	
Zinc	Dwarfism, hypogonadism	Metal fume fever
Polluting element		
Cadmium	—	Nephritis
Lead	—	Anaemia, encephalitis, neuritis
Mercury	—	Encephalitis, neuritis

with the sulphydryl groups of enzymes involved in the biosynthesis of haem and the cytochromes. *Figure 7* shows a typical input/output analysis for lead.

There are three types of poison based on mercury, namely, mercury vapour, inorganic (*ie* ionic) mercury and alkyl mercury. Mercury vapour can be absorbed in the lungs where, within a matter of hours, it is oxidised and so appears as ionic mercury in the bloodstream. These soluble inorganic mercury salts are extremely toxic mainly through corrosive action on the intestine, kidneys, and brain (for example, the phrase 'as mad as a hatter' refers to the mental disabilities caused by the use of mercuric nitrate in the felt hat manufacturing trade). However, an even greater danger from mercury arises from its alkyl derivatives. Such compounds are rapidly absorbed by erythrocytes (red blood cells) and then rapidly traverse the blood–brain barrier. This leads to disastrous, permanent injuries to the brain cells caused by the mercury compound becoming bound to proteins in cell membranes and so influencing the distribution of ions, electric potentials and passage of nutrients across these membranes. Another danger is that the mercury alkyl will cross the placental barrier and so cause serious damage to the foetus.

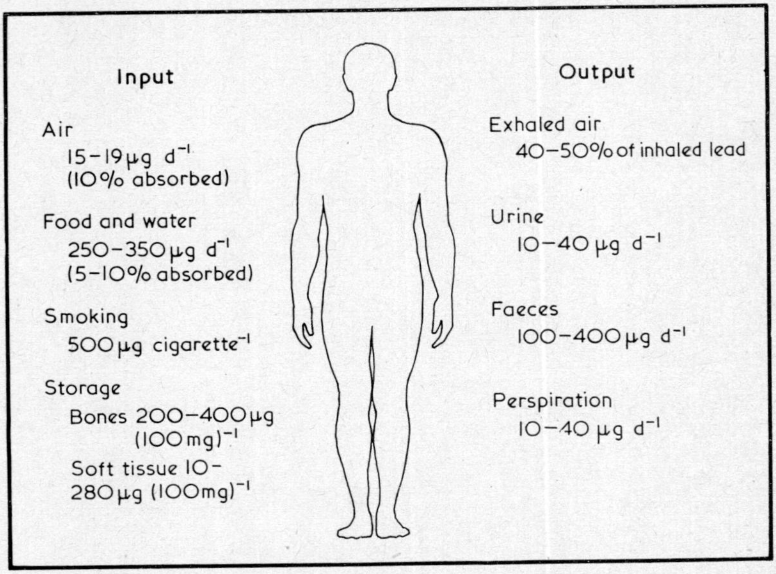

FIG. 7. Typical figures for the lead input and output for an American city dweller.

Cadmium occurs as 0.5 per cent of most zinc ores and so poisoning can arise from impure zinc plated sources. It causes renal damage, gastro-intestinal distress and cardiovascular malfunctions and it is carcinogenic. The chemistry of cadmium's disruptive action is one of challenging essential zinc ions.

Finally, we reiterate the principle that any species, be it element (essential, polluting *etc*) or ligand, taken into the body in excess is harmful no matter how important its presence is at lower concentrations (*see Fig. 8* and Table 3). Excessive accumulations may be a result of a metabolic malfunction, such as that of the mechanisms buffering the copper(II) concentrations in Wilson's disease (*see* Chapter 4), or of environmental or dietary origin (for example, correlations between the high calcium content of mains water and the low incidence of atherosclerosis or the chromium content of unrefined brown sugar and low blood cholesterol levels).

It would be foolish to declare all out war against modern technological processes which liberate polluting elements or sequester essential elements in the soil. Our wisest course is to try to understand and to maintain the balance of nature in which life on this planet has thrived for so long.

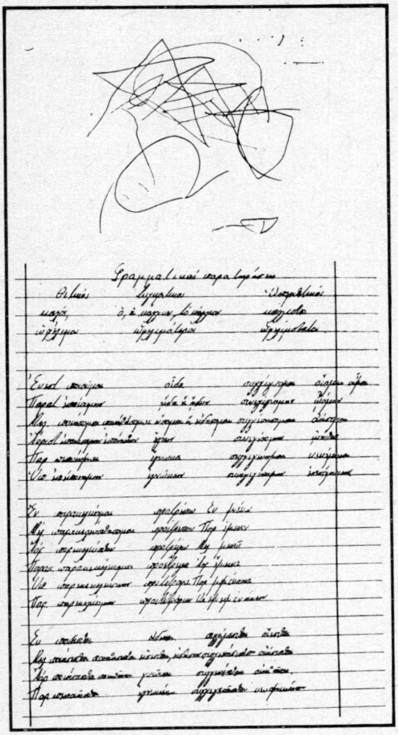

FIG. 8. Epilepsy resulting from a surplus of copper in the body. Handwriting of a patient (*a*) before and (*b*) after treatment with a copper sequestering drug. (Reproduced by kind permission of Professor J. M. Walshe.)

Trace element components of an ideal diet

The left-hand hump in *Fig. 2* has two sloping sides (A–B and C–D) which correspond to a deficiency, and to a surplus, of trace element respectively. Modern foods can combine to give diets which are deficient in some elements and/or have excesses of others.

Men have introduced non-food substances into their diets throughout recorded history, but, more recently, there has been an increase in the number of additives and in the commercial motivation for adding such substances—for flavour enhancement, colour improvement, extension of shelf life and protection of the nutritional value. Whatever one's opinion on additives, if food production is to keep pace with population growth, it is an established fact that chemicals not normally found in food, are going to play an increasingly important part. Although we may not be able to define an ideal diet exactly, if we keep in mind the basic bio-inorganic principles involved, we ought not to err too far from the straight and narrow.

Firstly, we must rehearse our vocabulary. 'Ingredients' are components normally present (but not necessarily in adequate amounts) in food whereas 'additives' are components not normally present. These latter may be ligands (which might sequester essential metals) or metal ions (of any of the four groups essential to polluting). One ought to consider the concentration of a metal and its reason for being present. There are often conflicting factors here, for example, a metal ion which improves the taste, and so appears to have its concentration optimised as far as the pleasure detecting centres of the brain are concerned, may, in fact, be at too high a concentration for the alimentary system or kidneys. Further, considerations of the health produced by adding an element and of the degree of safety (if, for example, a treble dose is administered) of its addition sometimes clash.

As mentioned earlier, no metal ion behaves in isolation—it is dependent upon both the other metal ions and the ligands present. Eggs, for example, contain iron as an ingredient and yet they actually suppress iron absorption in the small intestine. Sequestering ligands are sometimes used as additives to bind trace metals and hence prevent any oxidising action that these metals in an ionised state might catalyse in the food. Nature, through evolution, has chosen *homo sapiens* to survive because the competitive reactions in our bodies are able to cope with the ligands and metals in foods currently naturally available. This is not suggesting a return to 'health' foods. Under completely natural conditions *homo sapiens* could revert to being a naked ape brachiating through forests and constantly staring at extinction. If we were to return to the fruit, nuts, bulbs and roots diets of our ancestors more than 90 per cent of the human race would die of starvation. Man must have modern agricultural practices, fertilisers and additives in order to survive. However, just recently, there have been several challenges to nature's foods—(*i*) Simulated food products have come on the market, for example 'vegetable steaks', 'soya chicken breasts' and 'artificial fruit juices'. These are proteins, flavours, colourants, vitamins, and emulsifying agents but very few metal ions except perhaps sodium from monosodium glutamate which is used as a flavour enhancer. (*ii*) The receding glamour of the kitchen stove and the increasing number of working wives has caused a mushrooming in the sales of convenience foods with all their abundance of preservatives, colourants, flavourings, and texture agents. (*iii*) In some Western countries the young generation are being lured by pop idols to buy, and presumably eat, junk foods—those having neither the nutritive nor the energy values of natural food but being mere bulk and additives.

If the merchant guilds are to succeed in protecting the genuineness and reputations of their products, bio-inorganic chemists must begin to spell out ideal diets. Table 4 lists metal complexes that are currently permitted as additives (we have not listed the ligands because these run to well over a thousand). We could begin by analysing each metal compound's content in normal foods and seeing whether the new food vastly exceeds it. (It is not easy to establish norms for meat *etc* because all animal husbandry now involves trace element supplementation.) Secondly, we could examine idealised diets manufactured by pharmaceutical companies for feeding to patients requiring nutritional maintenance with a minimum of solid waste product excretion.

Table 4. Metal complexes that are currently permitted as food additives. Sodium and potassium salts are omitted for brevity.

Anticaking agents
- Aluminium calcium silicate
- Calcium silicate
- Magnesium silicate
- Sodium aluminosilicate
- Sodium calcium aluminosilicate
- Tricalcium silicate

Chemical preservatives
- Calcium ascorbate, propionate and sorbate
- Stannous chloride

Nutrients and dietary supplements
- Calcium carbonate, citrate, glycerophosphate, oxide, pantothenate, phosphate, pyrophosphate, and sulphate
- Copper gluconate, and iodide
- Ferric phosphate and pyrophosphate
- Ferrous gluconate, lactate and sulphate
- Iron, reduced
- Magnesium oxide, phosphate, sulphate, chloride, citrate, gluconate, glycerophosphate, hypophosphite, sulphate and oxide
- Zinc sulphate, gluconate, chloride, oxide and stearate

Sequestrants
- Calcium acetate, chloride, citrate, diacetate, gluconate, hexametaphosphate, monobasic phosphate and phytate

Stabiliser
- Calcium alginate

Miscellaneous additives
- Aluminium ammonium sulphate, potassium sulphate, sodium sulphate
- Calcium carbonate, chloride, citrate, gluconate, hydroxide, lactate, oxide and phosphate
- Magnesium carbonate, hydroxide, oxide and stearate

These chemically defined elemental diets were first used in the 1930s, various formulations being composed to study the physiological effects of variations in specific nutrients. The last decade has seen a vast expansion in the use of such diets as a means of sustaining astronauts with the minimum of defecation, as element sterile diets for experimental animals in order to designate the excluded element as 'essential,' 'beneficial' *etc*, and as a means of supporting humans who, for medical reasons, cannot accept normal food (for example, persons with malabsorption states, short bowel syndrome, total gastrectomy, bowel obstruction or fistula). These diets have reached a remarkable level of sophistication. For example, 15 normal adult males have been kept healthy on such diets for more than five *months*. Current work is aimed at improving the palatability of these chemically defined diets, at searching for micro-trace elements (*eg* chromium, molybdenum, nickel, selenium or vanadium) that treatment over many *years* ought to include, and at including the correct ligands as counter ions to the metals in order to ensure that the elements are in their most readily assimilated forms

Table 5. A typical element content of an artificial nutritional diet. The salts are mixed with vitamins, amino acids, carbohydrates.

Cations	mg d^{-1}	Anions	mg d^{-1}
Sodium	1548	Chloride	3246
Potassium	2105	Phosphate	500
Calcium	800	Sulphate	11.5
Magnesium	350	Iodide	0.144
Manganese	2.81		
Iron	10.0		
Copper	1.94		
Zinc	12.5		
Cobalt as Vit B_{12}	0.22 μg		

(for example, *see Figs 42* and *43*). A typical element content of one of these diets is shown in Table 5.

How do people get perfect nutrition—every item being in just the right amounts? The current answer is that they do not. People survive on very imperfect nutrition and just as an undernourished field of corn does not produce a maximum yield of grain so too our efficiency is not all it might be. It is well within the capability of modern science to tackle the problem of producing sufficient components for better diets for man and even the more difficult problem of exactly what such diets ought to contain.

3. The Roles of Inorganic Species *in Vivo*

In 1946 The World Health Organisation defined 'health' as 'a state of complete physical, mental and social wellbeing and not merely the absence of disease or infirmity'. As we have seen in the previous chapter, at least 26 elements are necessary for man's achievement of this state of health. Not only must these elements be present in the body but also they must be in the correct locations, in the correct amounts, in the correct oxidation states and bound to the correct chemical partners.

Concentrations of even simple ionic species vary widely between different bodily locations, *eg* intracellular and extracellular fluids (*see Fig. 9*).

The elements of organic matter H, C, N, O, P and S constitute by far the greatest proportion of the body's atoms, *eg* 63 per cent are H, 25.5 per cent are O, 9.5 per cent are C, 1.4 per cent are N and the other 20 elements account for less than 0.7 per cent. The aforementioned six elements compose all of the body's organic molecules, *eg* proteins, carbohydrates, fats, nucleotides and the energy transfer molecules ATP and ADP. The functions of these species have been discussed adequately elsewhere and we shall not consider them further (*eg* see Brown's Monograph).

The remaining biological elements can be classified as main group elements: Na, K, Mg, Ca and Cl which are usually mobile as ions, or trace elements: Zn, Fe, Cu, Ni, Mo, Co, Mn, Cr, V, Si, Sn, Se, F, Br and I which usually have fixed chemical neighbours in a biological system.

Main group ions

It is difficult to investigate main group metal ions *in vivo* because they are not coloured, are not paramagnetic and do not take part in oxidation–reduction reactions. However, several general functions have been identified.

The ionic species Na^+, K^+, Ca^{2+}, Mg^{2+}, Cl^-, SO_4^{2-} and PO_4^{3-} maintain the balance of charges in body fluids and cells and maintain the proper liquid volume of blood and the other fluid systems. As can be seen from *Fig. 9* the distribution of these ions is highly specific *eg* K^+ and Mg^{2+} inside cells and Na^+ and Ca^{2+} in plasma. This stabilises the cell against osmosis, by rejection of Na^+, and against internal precipitation of carbonate and phosphate by rejection of Ca^{2+}. This discrimination between intracellular and extracellular composition is maintained by extremely complicated multiple enzyme membrane processes called 'ion-pumps'. When there is such an

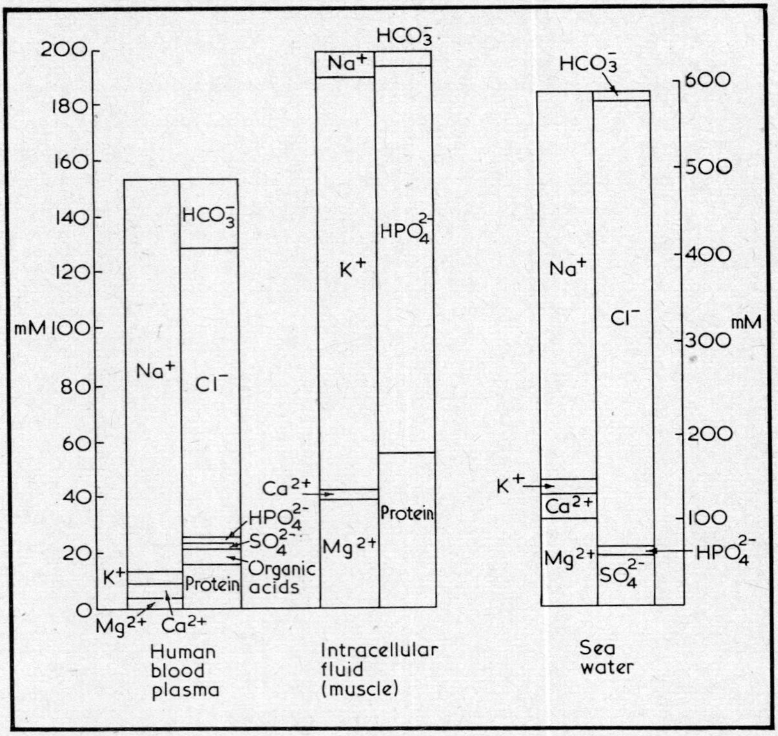

FIG. 9. The solute composition of some human body fluids compared with sea water.

imbalance of ion concentrations across a membrane the natural tendency is for ions to flow down the concentration gradient until the concentrations are equalised. However, 'ion-pumps', which are active transport processes, work against this tendency and maintain the constant optimal internal concentrations of inorganic electrolytes which are essential for the function of important intracellular processes (*see Fig. 10*).

Sodium and potassium

Na^+ and K^+ have closed shell ionic structures and so electrostatic forces rather than covalent bond formation are important in their activity. Which complexes are formed depend upon the ionic charges and radii, and they are entropy stabilised.

Gradients of Na^+ and K^+ across cell membranes are primarily responsible for the transmembrane potential difference which, in

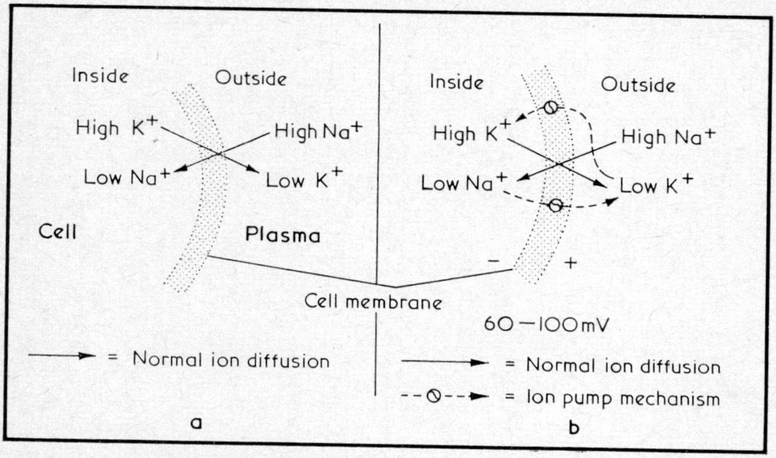

FIG. 10. The mechanism of ion transport across cell membranes. (*a*) Passive diffusion, and (*b*) involving an ion pump mechanism.

nerve and muscle cells, is responsible for the transmission of nerve impulses.

Sodium has few specific functions, its general role being in maintaining osmotic pressures and membrane potentials. However, the extrusion of Na^+ from a cell is involved in the active transport of amino acids and of sugars into the cell.

Potassium has the lowest charge density of the four main group ions and so has the possibility of diffusing through hydrophobic solutions, *eg* lipid-protein cell membranes, almost as easily as it diffuses through water. K^+, being one of the most important intracellular cations, acts as a cofactor for some internal enzymes and stabilises internal structures. Pyruvate kinase, for example, which is essential for glycolysis, requires a high K^+ concentration and is inhibited by Ca^{2+} and Na^+. The most critical life process is protein synthesis by the ribosomes and these require a high K^+ concentration for maximal activity.

Calcium and magnesium

Calcium and magnesium play a variety of important structural and catalytic roles in overall cellular metabolism and so have evolved as intrinsic ingredients in the complex mass of metabolites found in all cell systems. Like Na^+ and K^+ these ions help to maintain membrane potentials and transmit nerve signals. These two metals bridge neighbouring carboxylate groups in lipoproteins and so stiffen cell membranes. In fact, in the absence of calcium, cell membranes

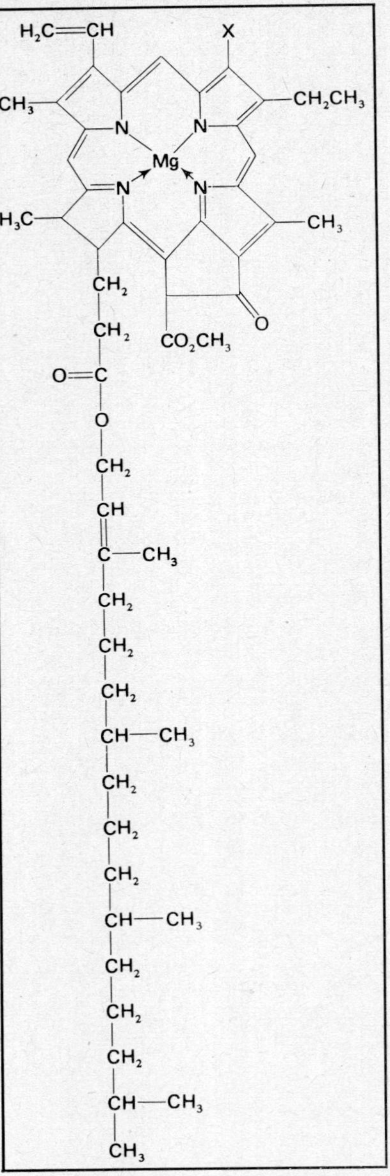

FIG. 11. The structure of chlorophylls. In chlorophyll a, X = $-CH_3$; in chlorophyll b, X = $-CHO$.

become porous. Generally, these ions act as catalysts at weaker base centres such as the polyphosphates.

As we have seen calcium is mainly extracellular and so it acts as an external structural factor and also as a cofactor for extracellular enzymes. The most important role of calcium in the body is, of course, the formation of bone, hydroxyapatite. There is a very fine control over the precipitation of calcium salts and so bone and shell material can be transferred in the bloodstream to be deposited in a new region. This metal also takes part in numerous vital physiological processes such as blood-clotting, hormone release, lactation, nerve conduction, muscle contraction and the formation of intercellular 'cement'. Ca^{2+} also antagonises the activating effect of Mg^{2+} in many enzymic processes and so is very useful in enzyme control mechanisms.

Magnesium, on the other hand, is a stabiliser of internal structure and a cofactor for intracellular enzymes. Nucleotides inside cells exist as their Mg^{2+} complexes, since Mg^{2+} binds preferentially to phosphates, and so magnesium is necessary for DNA replication and protein biosynthesis.

A very important function of magnesium is in photosynthesis. Solar energy is the ultimate source of all biological energy and photosynthesis, being the process by which this energy is converted to chemical energy, is of prime importance. All photosynthetic cells carry out the reaction

$$nH_2O + nCO_2 \xrightarrow{\text{light}} (CH_2O)_n + nO_2$$

which is dependent upon one of the green pigments known as chlorophylls which contain Mg^{2+} as their central metal ion (*see Fig. 11*).

Trace elements

In many cases the exact roles of the trace elements are still unknown. However, the great majority of these serve as key components of essential enzyme systems or of other proteins which perform vital functions.

There are two types of enzyme which require metal ions (*i*) metallo-enzymes: in these the metal ion is firmly bound to and is an integral part of the enzyme protein molecule; or (*ii*) metal ion activated enzymes: in these the metal ion is bound weakly and several different metals may activate the same enzyme.

These enzymes may be very complicated, *eg* xanthine oxidase, which catalyses the reaction,

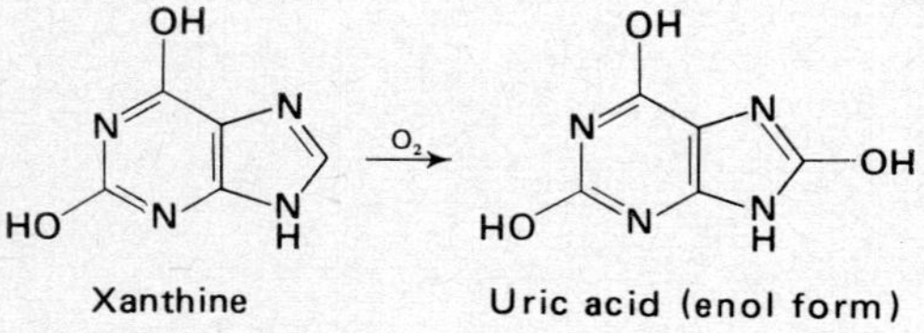

has a molecular weight of 300 000 and contains eight iron and two molybdenum atoms per molecule.

The function of the metal ion in these systems may be of several types:

- to lock the geometry of the enzyme protein so that only a specific substrate can be attached
- to bind the substrate to the protein and so act as a template, *eg* the cyclisation of the corrin ring can use Ni(II) as a template
- to redox in a reaction in which the substrate is either oxidised or reduced.

Other proteins which involve metal ions are the transport and storage proteins which may control the concentrations of either a metal or an important substrate such as oxygen or carbon dioxide.

We shall consider the biological roles of each element in turn, noting that these are always chemically sensible and that the more common species perform the more common tasks, *eg* Fe and Cu are used for oxygen transport and assimilation.

Zinc

Zinc, which was shown, in 1934, to be essential for the normal growth and development of mammals, is present in the human body to the extent of 1.4 to 2.3 g. This metal has a full 3d shell of electrons and so is not strictly a transition metal, but it does form transition metal-like complexes and is a strong Lewis acid. In aqueous solution there is no evidence for zinc in any oxidation state other than +II.

Zinc complexes form good pH buffering systems *in vitro* and are used in the highly important pH control mechanism *in vivo*. The blood plasma of man has a normal pH of 7.4 and if this falls below 7.0 or rises above 7.8 irreparable damage may occur, enzymes being particularly sensitive to pH changes. Since a variation in hydrogen ion concentration of 3×10^{-8} mol dm^{-3} can be lethal we must have a very sensitive control system governing its concentration.

In the body there are about 18 zinc metalloenzymes and approximately 14 zinc ion activated enzymes. Zinc, being a strong Lewis acid, prefers to complex to phosphates and so is involved in their hydrolysis by phosphatases which is important in RNA synthesis. Other enzymes incorporating this metal help control the rate of formation of CO_2 from HCO_3^- and the digestion of proteins. There are also complicated interactions between the metabolisms of zinc, copper and iron, *eg* zinc affects the incorporation into and the release of iron from ferritin, an iron storage protein.

Iron

Iron is the most abundant transition metal in the human body, 4.2 to 6.1 g in the average man, and is involved in a great variety of metabolically active molecules.

In vitro, in solution, iron can redox between Fe(II) and Fe(III), the ferrous state being easily air oxidised to ferric except in acid solution. Both of these oxidation states are strong Lewis acids and so can accept electrons, to bond to oxygen, for example.

In vivo iron may be held in either of its oxidation states or may redox between the two, depending on the ligand to which it is attached.

The oxygen transport protein of red blood cells, which transfers oxygen from the lungs to the tissues, is haemoglobin. The solubility of O_2 in water is only 0.5 cm^3/100 cm^3 but, when haemoglobin is present in physiological concentrations, this is increased to 20 cm^3/100 cm^3 of whole blood. In haemoglobin and in myoglobin, the oxygen binding protein of muscle, iron is held in the +II oxidation

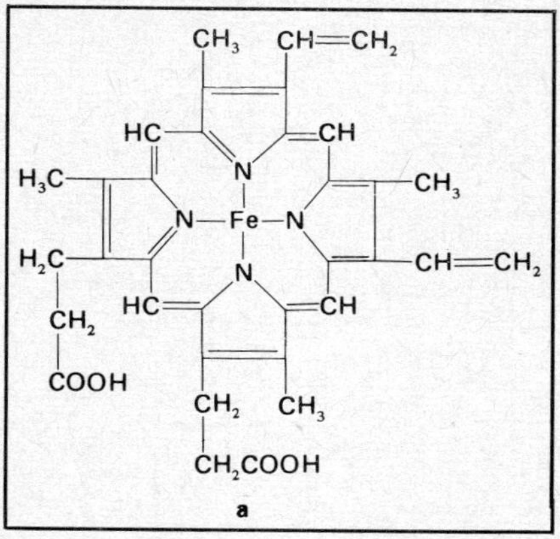

FIG. 12. Structure of myoglobin. (*a*) Protoporphy in IX, a tetrapyrrole ring. (*b*) Complete myogiobin structure MW 17 000.

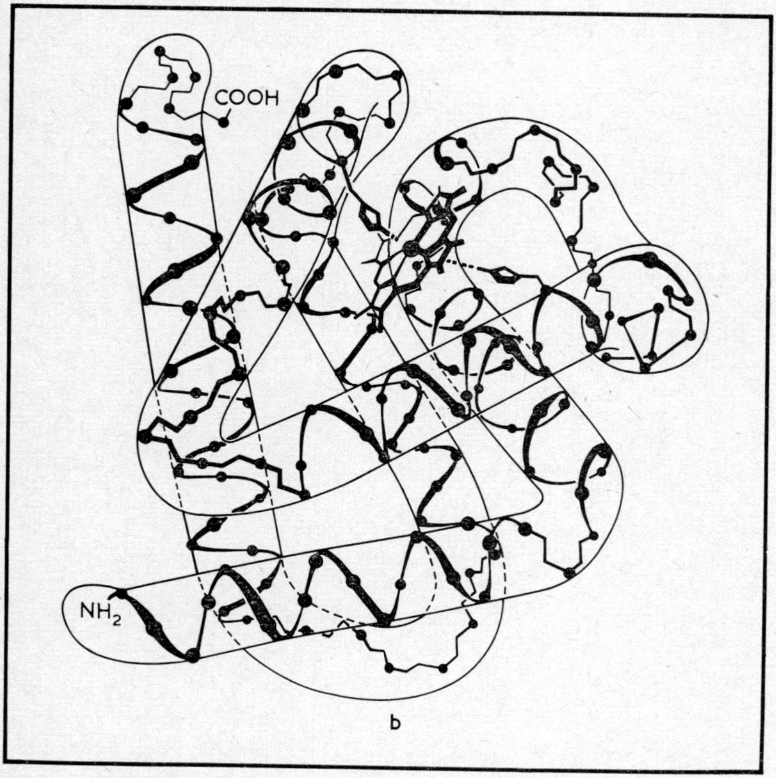

state in the centre of a tetrapyrrole ring which is incorporated into a protein molecule (*see Figs 12a* and *b*).

The normal coordination number of Fe(II) is six. As we have seen, four of these positions are taken up by nitrogen atoms. The fifth is occupied by an imidazole nitrogen atom from a histidine residue on the protein moiety and the sixth may reversibly attach O_2. The protein environment of the iron atom is arranged so that Fe(II) is the preferred oxidation state thus allowing O_2 to be bound to the iron without oxidising it.

In the catalases, which are very large molecules (MW 247 500), and oxidases, iron is held in the +III oxidation state. The role of catalase in biological oxidations is not known with certainty but it is believed to catalyse the decomposition of hydrogen peroxide which is produced in some cell microbodies.

In the molecules in which iron redoxes, the actual oxidation–reduction potential is directly related to the affinity of the ligand environment for the two oxidation states in question. For example, the cytochromes, which are important components of the mitochondrial electron transport chain, are a group of proteins which transfer electrons from flavoproteins to molecular oxygen. They all contain iron–porphyrin prosthetic groups similar to haemoglobin and myoglobin but in this case Fe(II–III) valence changes occur.

Copper

Copper is known in two oxidation states in solution, Cu(I) being easily oxidised to Cu(II) unless it is complexed. Thus, as we would expect, copper's main role *in vivo* is in redox reactions. This metal is present in about 12 enzymes whose functions range from the utilisation of iron to the pigmentation of the skin.

Copper is necessary for the release of iron from its storage protein and ceruloplasmin is believed to be active in the oxidation of ferrous iron. So copper is essential for iron metabolism and oxygen transport in vertebrates but in some invertebrates a blue copper protein, haemocyanin, is actually used to carry oxygen.

The two copper containing enzymes, amine oxidase and tyrosinase are of importance, the first for the formation of elastin and bone collagen and the second for the formation of melanin, the skin colouring pigment. Genetic deficiency of tyrosinase results in albinism.

As mentioned, the cytochromes of the electron carrier chain contain iron, however, the last enzyme of the chain, cytochrome oxidase, contains a copper atom as well as the iron-porphyrin group. This is the only one of the cytochromes which can be reoxidised by molecular oxygen and it is known that during the electron transport to oxygen Cu(I–II) transitions occur.

Molybdenum

In vitro the main oxidation states known for molybdenum are V and VI, III and IV also being known but always in the form of complex species and always being air-sensitive. Molybdenum can coordinate to a variety of different ligands and participates in both redox and ligand exchange reactions *via* a number of different mechanisms.

The major routes of nitrogen incorporation into plants, and hence into man, are nitrate reduction and nitrogen fixation. The enzymes for both of these vital processes require molybdenum as well as iron. The former metal is also necessary for the activity of flavoproteins, xanthine oxidase and aldehyde dehydrogenase but there does not appear to be a specific molybdenum-carrying protein.

Nickel

Very recently nickel has been shown to be essential for the growth of rats but little is yet known about its role *in vivo*. However, nickel deficiency does impair iron absorption and leads to reduced iron content in the organs, and to reduced haemoglobin levels and red cell counts.

Cobalt

Cobalt, which has been known to be an essential element since the 1930s, is unique in that it is an essential component of a vitamin, vitamin B_{12}.

Blood contains only 2×10^{-4} μg of vitamin B_{12} per cm^3 and many fruitless attempts were made before it was eventually isolated in 1948.

The ultimate source of all vitamin B_{12} is synthesis by microorganisms. The microflora of the human intestinal tract do perform this synthesis but the vitamin thus formed cannot be reabsorbed into the body and so humans are entirely dependent on their diet as a source of this vitamin.

It took many years for the complicated structure of vitamin B_{12} to be elucidated, *see Fig. 13*. Vitamin B_{12} and its derivatives are involved in a great many biological reactions mainly in the synthesis of DNA and haemoglobin, the metabolism of amino acids, hydride transfer and methyl group transfer. It is proposed that during these reactions Co(II) is reduced to Co(I).

The cobalt ion itself is believed to be an essential cofactor for some enzymes but of these only glycylglycine depeptidase is known to occur in animals.

Manganese

Manganese is known in eight different oxidation states *in vitro* but *in vivo* only +II and +III occur. Even though +III is unstable *in*

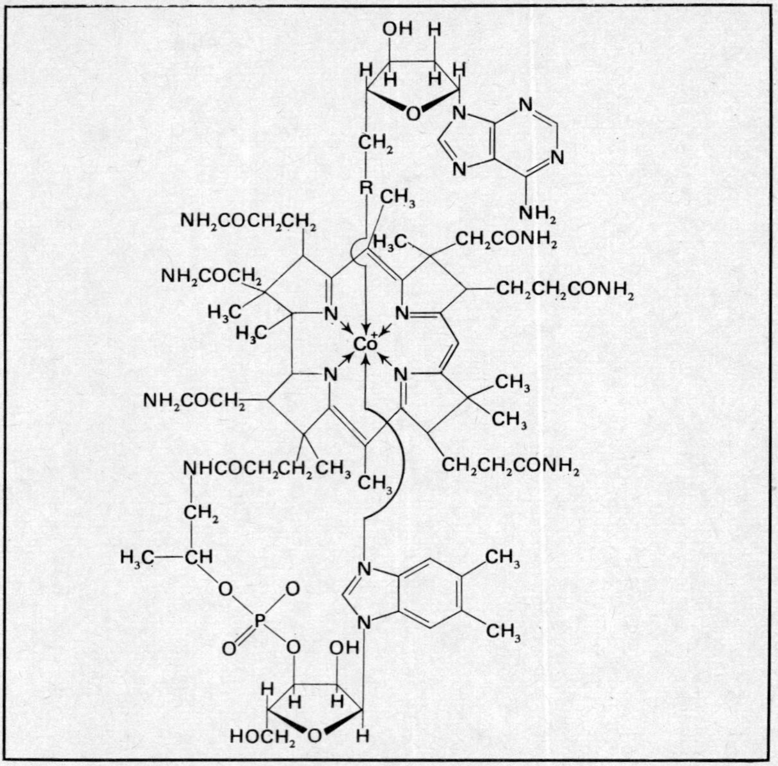

FIG. 13. The structure of Vitamin B_{12}.

vitro it is believed to be the form present in the manganese transporting protein, transmanganin. This metal accumulates in, and is necessary for the function of, the mitochondria and so it is proposed that it is a cofactor for the respiratory enzymes which are situated there.

Mn(II) shows chemical similarities to Mg(II), *eg* it binds preferentially to weaker donor ligands such as carboxylate and phosphate. In fact, in reactions involving phosphate, Mn(II) can replace Mg(II). For example, Mn–ATP complexes are important in some reactions. Only two Mn metalloenzymes are known, arginase and pyruvate carboxylase, but there are also several Mn ion activated enzymes, *eg* iso-citric dehydrogenase, polymerases and galactotransferases.

Chromium

Chromium is essential only for the health of higher animals and was not shown to be necessary until 1955. It works in conjunction with insulin, and experiments in humans have confirmed that it is involved in carbohydrate utilisation. The physiologically active chromium complex has not yet been identified but it is known to be constantly present in nucleic acids where it is probably in the Cr(III) form bound to phosphate groupings. As well as influencing nucleic acid synthesis, this metal also affects lipid and cholesterol synthesis and amino acid utilisation. Physiological amounts of Cr are transported in the blood attached to plasma proteins including transferrin on which Cr(III) competes for binding sites with iron.

Vanadium

Although vanadium has been shown to be beneficial to man its function is not known. There is some evidence that it is involved in lipid metabolism and it may prevent dental caries.

The sedentary marine animal *Ascidiella aspersa* lives on the floor of the Bay of Sebastopol in the Black Sea and contains vanadium in quantities which amount to 400 kg per square mile of sea floor. This vanadium is in the low valency state +III which requires high acidity (equivalent to 9 per cent sulphuric acid!) to prevent oxidation. The vanadium in these animals is used in oxygen transport, the transporting molecule being haemovanadin, and is also involved in building the cellulose like molecules of the substance which envelops them. In some marine worms vanadium may replace iron or copper as respiratory pigments.

Silicon

Silicon plays an important structural role *in vivo*, influencing the biosynthesis of collagen and bony tissues. It is essential for the normal functioning of epithelial and connective tissue, to which it imparts strength and elasticity by holding the protein molecules together *via* crossbridges similar to those formed by sulphur. This crosslinking in skin gives it greater chemical and mechanical stability and renders it impermeable to liquids. The permeability and elasticity of blood vessel walls are also determined by the presence of silicon.

Tin

Tin would appear to be beneficial to mammals but data on its *in vivo* functions is very limited. Certainly, its chemical properties would suggest that a biological function is possible, *eg* $Sn^{2+} \rightleftharpoons Sn^{4+} + 2e^-$ can occur, Sn can be four, five, six, or eight coordinate and it has a strong tendency to form covalent links to carbon, a great many organic-tin compounds being known.

One postulated role of tin is in the stimulation of the burst of

DNA/RNA transcription and translation and protein synthesis which happens shortly after birth.

Selenium

As mentioned in the previous chapter, the biological effect of selenium is highly dependent upon its concentration, *ie* 0.1 ppm is beneficial whereas 10 ppm is carcinogenic. The metabolic role of selenium is not known but it is thought to be a part of the enzyme glutathione peroxidase which protects haemoglobin against the oxidative effects of hydrogen peroxide.

The biologically active form of selenium seems to be the selenide, R_2Se, which is protected from oxidation by vitamin E. This selenide is believed to form part of a non-haem iron protein which is involved in electron transfer in mitochondria and the smooth endoplasmic reticulum.

Fluorine, chlorine, bromine, iodine

Two of the halogens, chlorine and iodine, are essential but the other two are not.

Iodine is heaviest of the essential elements being a necessary constituent of the thyroid hormones, thyroxine and triiodothyronine, but its precise hormonal activity is still not understood.

Fluorine, as we must all have heard, is beneficial for the formation of sound tooth and bone tissue.

All animal tissues, except thyroid, contain 50 to 100 times more bromine than iodine but bromine has not yet been shown to perform any vital function in higher plants or animals.

Solvents in vivo

When the first cells evolved in the oceans they must have been capable of tolerating their aqueous environment. Through evolution modern cells have been produced which use the unique properties of water to their best advantage. Water is the only naturally occurring inorganic liquid and also the only compound that occurs in nature in all three states of matter—as a solid, a liquid and a gas. The omnipotence of the roles of water in humans may be seen by reference to Table 6. The liquid is not only used to supply bulk to the body but use is also made of its unusual chemical properties.

Water has an exceptionally high specific heat and so it is able to act as a heat buffer thus minimising the effect on the cells of any environmental temperature fluctuations. Because water has a high latent heat of vaporisation, vertebrates can use the evaporation of sweat as a cooling mechanism.

The hydrogen bonding between molecules of liquid water means that it also has a high melting point, boiling point, heat of fusion and

Table 6. Some features of water in humans. These figures are adapted from those given by F. Franks, *Chem. Brit.*, 1976, **10**, 278.

Percentage composition	%
Adult human	65–70
Human embryo, 1st month	93
Nervous tissue	84
Liver	73
Muscle	77
Skin	71
Connective tissue	60
Adipose tissue	30
Plasma, saliva, gastric juice	90–99
Adult input in cm^3 per day	
Drink	1300
Food	850
Food oxidation	350
Adult output in cm^3 per day	
Respiration	400
Evaporation	500
Excretion	1600
Adult turnover in cm^3 per day	
Saliva	1000–1500
Gastric juice	1000–2000
Bile	500–1000
Pancreatic juice	600–800
Intestinal juice	200
Adult throughput in litres per day	
Kidney	180
Heart	7000

surface tension. Hydrogen bonding is also important in biological molecules other than water, for example, the two spirals of the DNA double helix are held together by hydrogen bonding between complementary base pairs, *see Fig. 14*.

Many important properties of macromolecules depend upon their interactions with water molecules.

Water, being dipolar in nature, is a very good solvent for a variety of solids. Most ionic compounds, of course, dissolve because the ion solvation provides enough energy to break the crystal lattice

FIG. 14. The complementary base pairs of DNA.

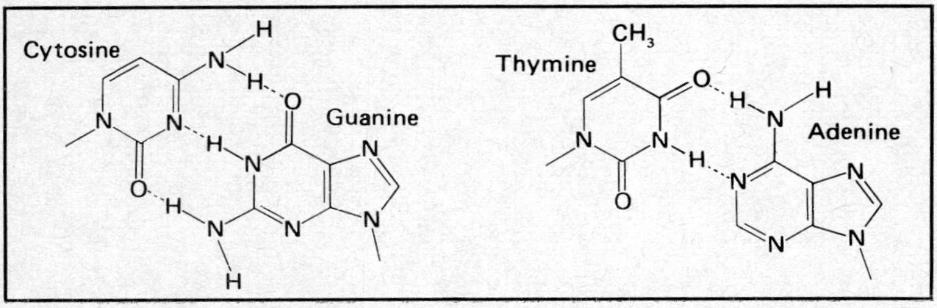

structure. Polar compounds, for example, alcohols, aldehydes and ketones, dissolve due to the ability of water to form hydrogen bonds to their polar groups. Molecules containing both hydrophilic and hydrophobic groups, for example, fatty acids and polar lipids, are dispersed in water as micelles.

The control of the degree of ionisation of water is also of vital importance, the main buffering reactions being

$$H_2PO_4^- \rightleftharpoons HPO_4^{2-} + H^+ \text{ inside cells and}$$
$$H_2CO_3 \rightleftharpoons HCO_3^- + H^+ \text{ outside cells.}$$

The membranes of biological systems and the active sites of enzymes may be considered as special solvents in their own right. In membranes, lipids, proteins and carbohydrates make up a system through which ionic and non-ionic species must pass, in some cases by simple diffusion and in others by specific active transport processes. In the active sites of enzymes, exceptional conditions exist in which substrates may be held in stereochemistries which would be extremely unstable *in aquo*, thus modifying their chemical activities.

4. Principles—at the Molecular Level

As long ago as 1780 Lavoisier described life as a chemical process but, even today, only a little is known about the life process at the molecular level.

As was discussed in Chapter 3 the control of the pH of a system is achieved by the use of a buffer system, both intracellular and extracellular fluids of living organisms containing conjugate acid–base pairs which act as buffers at the normal pHs of these fluids. However, the hydrogen ion is by no means the only species to have its concentration held constant *in vivo*. In fact, metal ions and ligands are also homeostatically controlled (*ie* their concentrations are buffered) at each stage of their cycle in the organism and also in their interaction with the environment.

In the mammalian body the concentration of ligands is much greater than that of metals and so various complexing species compete for each metal ion. Also hydrogen ion is competing with the metal ions for each ligand. In *Fig. 15* is shown the simple *in vitro* interaction between zinc(II) and aspartic acid. As can be seen from the concentrations of species present, at pH 7 for example, halving the metal concentration does not simply halve the concentration of each complex. *In vivo* the situation is much more involved, the homeostatic level of any metal depending on many factors, for example

(*i*) the species, age and sex of the animal,
(*ii*) the organ within the animal,
(*iii*) the time of day,
(*iv*) the diet of the animal,
(*v*) the presence of pathological conditions,
(*vi*) the pH of the organ, and
(*vii*) antagonism or stimulation by other metals.

Let us consider the homeostasis of iron in man as an example. Iron metabolism may be represented as in *Fig. 16*. The normal adult body contains 4.2–6.1 g of iron, 65–70 per cent of which is present as haemoglobin in the red blood cells. The average red cell life is about 120 days and, since haemoglobin levels are held constant, about 25–30 mg of iron per day is necessary for haemoglobin synthesis. This synthesis occurs in the bone marrow and so iron, as its transferrin complex, has to be transported through the bloodstream from its sites of storage. The formation of this Fe(III) transferrin

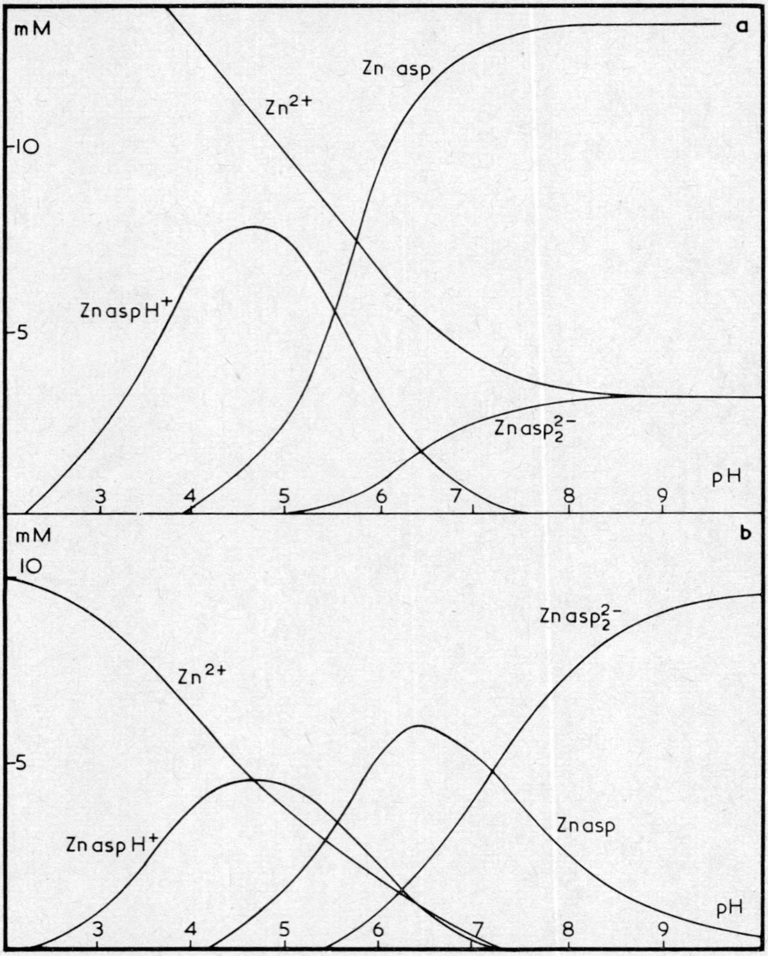

FIG. 15. Species present due to the interaction of zinc(II) and aspartic acid.
(*a*) concentration of aspartic acid $= 2 \times 10^{-2}$ mol dm^{-3}
concentration of zinc(II) $= 2 \times 10^{-2}$ mol dm^{-3}
(*b*) concentration of aspartic acid $= 2 \times 10^{-2}$ mol dm^{-3}
concentration of zinc(II) $= 1 \times 10^{-2}$ mol dm^{-3} asp = (aspartate)

complex depends upon factors such as the pH and the presence of CO_2 or HCO_3^-.

The average adult male has a serum iron level of 130 μg/100 cm^3 but the following variations may occur in humans.

(*a*). The level in the female is 10–15 per cent lower.

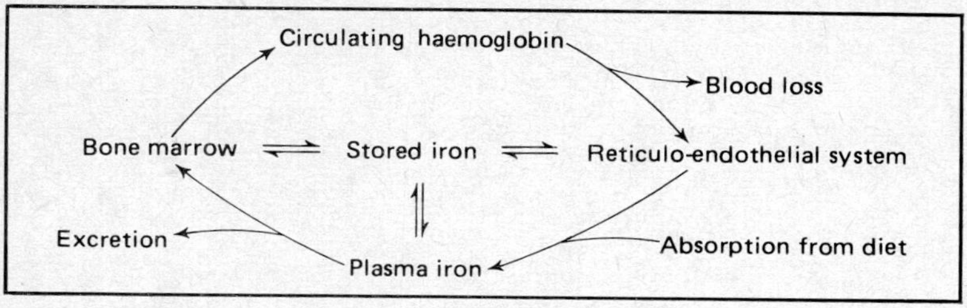

FIG. 16. Iron metabolism in man.

(*b*). Plasma iron is highest in the early morning and lowest during the afternoon.
(*c*). The plasma iron level gradually falls during pregnancy.
(*d*). At birth the plasma iron level is about 150–200 μg/100 cm^3 but within a few hours this falls to less than 100 μg/100 cm^3 and then slowly and irregularly rises to reach adult levels between the ages of three and seven years.
(*e*). Dietary deficiencies, such as protein, vitamin C or vitamin B_{12} deficiency, produce iron deficiency. Excessive iron intake, *eg* from iron cooking pots, causes iron to be deposited in some organs.
(*f*). During pathological conditions, such as liver disease, marrow disease, infection or inflammation, the iron level may be drastically altered.

Thus, we see that to unravel the chemistry of such complicated processes will require a great deal of study and indeed a battery of experimental techniques have been brought to bear on the problem. In fact almost every analytical technique from organic, inorganic or physical chemistry has been adapted and applied to biochemical systems. If we can isolate a component from the living system then we have the means of determining its molecular weight, atomic composition, size, shape, bond strengths and the oxidation numbers of any metals present. Methods applied are, for example, nmr spectroscopy, mass spectroscopy, esr spectroscopy, light scattering techniques, x-ray crystallography, microcalorimetry and, of course, standard organic methods of degradation and synthesis. The reactions of biological molecules are often very rapid and have to be studied by special methods such as stopped flow relaxation techniques.

However, one must always be careful when extrapolating results from an isolated *in vitro* system to the *in vivo* situation—crystal structures determined by x-ray spectroscopy are not necessarily the structures found in solution in the bloodstream.

It is by no means always possible to discover the desired information by isolating components from a system and the complete living unit may have to be studied. Techniques for achieving this are still in their infancy and are seldom applied to species any more complicated than bacteria.

Simple ligand–metal ion complexes in vivo

The principle of selectivity of chemical reactions is well known to the inorganic chemist, for example, it enables one to detect a particular metal ion in the presence of several others. This individuality of each metal can explain to a large extent their roles in the bio-system.

'Specificity' of a ligand for a single metal ion is very rare; even enzymes can generally have their active metal ions replaced by others, often leading to a loss of activity but occasionally giving a more active species. However, a much higher degree of 'selectivity' is found *in vivo* than can be reproduced *in vitro*.

Nature is very thrifty and so many ligands are used for more than one purpose and may require to interact with several different metals. In this case, of course, complete specificity of the ligand would be undesirable. As can be seen from the structures given in Chapter 3, chlorophyll, myoglobin and vitamin B_{12} all have structures depending on a similar ligand but using magnesium, iron and cobalt respectively.

The theory of hard and soft acids and bases

Since the 1940s coordination chemists have been aware of trends in formation constants of metal complexes, for example, the Irving–Williams series of stability: for a given ligand the formation constants of a complex with a divalent metal ion are in the order

$$Ba^{2+} < Sr^{2+} < Ca^{2+} < Mg^{2+} < Mn^{2+} < Fe^{2+} < Co^{2+} < Ni^{2+} < Cu^{2+} > Zn^{2+}$$

The complexing power of a metal ion depends on its charge to radius ratio and, in the case of the transition metal ions, on its ligand field stabilisation energy. The effect of these factors can be seen from *Fig. 17*.

Solution chemists have built up a vast amount of experimental data and from these the principles of modern coordination chemistry have been deduced. The systematic arrangement of chemical facts was applied to complex solution chemistry in the 1950s by Ahrland, Chatt, Davies and Schwarzenbach who noticed that different cations behaved very differently when offered a series of ligands. For example, certain ligands were seen to form most stable complexes with Ag^{+}, Hg^{2+} or Pt^{2+} whereas others preferred Al^{3+}, Ti^{4+} or Co^{3+}. Thus, many years of observation were used to divide metal ions, in aqueous solution, into two classes (*a*) and (*b*) depending on their

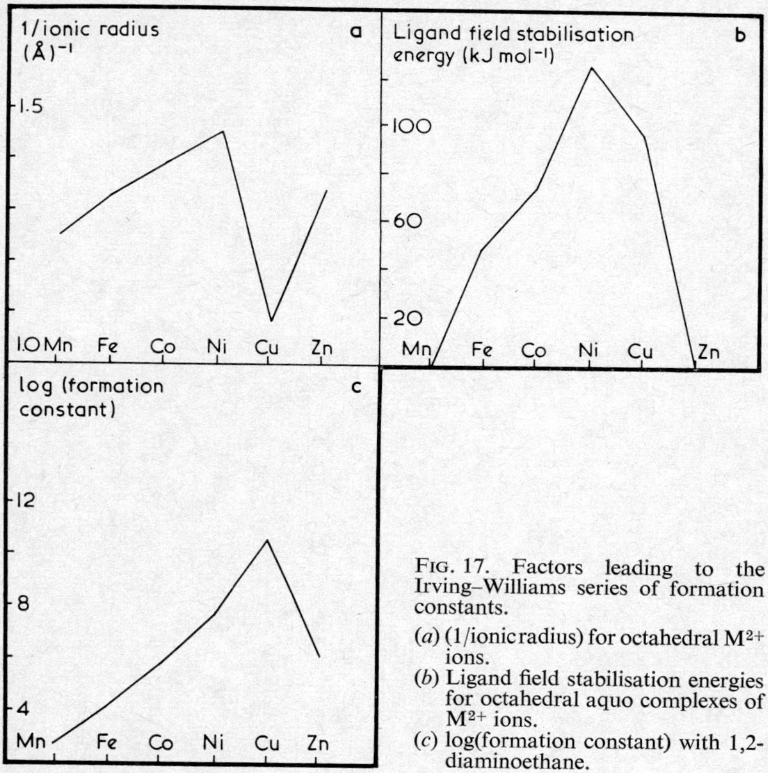

FIG. 17. Factors leading to the Irving–Williams series of formation constants.
(*a*) (1/ionic radius) for octahedral M^{2+} ions.
(*b*) Ligand field stabilisation energies for octahedral aquo complexes of M^{2+} ions.
(*c*) log(formation constant) with 1,2-diaminoethane.

behaviour towards complexing atoms. Unfortunately, the division is not sharp and a great many species fall into the category 'borderline'. Part of this classification is given in Table 7.

Table 7. Classification of some cations.

(*a*)	H^+, Li^+, Na^+, K^+, Mg^{2+}, Ca^{2+}, Mn^{2+}, Cr^{3+}, Fe^{3+}, Co^{3+}
Borderline	Zn^{2+}, Cu^{2+}, Ni^{2+}, Fe^{2+}, Co^{2+}, Sn^{2+}, Pb^{2+}
(*b*)	Cu^+, Ag^+, Au^+, Tl^+, Pd^{2+}, Pt^{2+}, Cd^{2+}

In 1963, Pearson used the terms 'hard' and 'soft' to describe class (*a*) and (*b*) metals respectively. He considered all complexation reactions as being between an acid and a base part according to the Lewis definition of an acid as an electron acceptor and a base as an electron donor. It is important that the distinction between Lewis acidity and inherent acid strength is understood, *eg* OH^- and F^- are both 'hard' bases but the basicity of OH^- is about 10^{13} times that of F^-. Bases also could be classified as 'hard' or 'soft', *see* Table 8.

The empirical rule was then expounded that "'hard' acids prefer to bind to 'hard' bases and 'soft' acids prefer to bind to 'soft' bases". This is an extremely useful concept which can be used to predict the strengths of complex bonds.

Table 8. Classification of some anions.

(*a*)	H_2O, OH^-, ROH, OR^-, R_2O, NH_3, $-NCS^-$, Cl^-, PO_4^{3-}, SO_4^{2-}, F^-, NO_3^-, CO_3^{2-}
Borderline	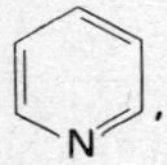, RNH_2, N_2, N_3^-, NO_2^-, Br^-
(*b*)	RSH, RS^-, R_2S, R_3P, R_3As, CO, $-CN^-$, $-SCN^-$, $S_2O_3^{2-}$, H^-, I^-

R = alkyl group

The characteristics of the two classes are as follows:

Class (*a*)—these metals generally have a rare gas octet structure, *eg* K^+. Their interactions are primarily electrostatic with the bond strengths depending on the charge to radius ratio. However, the size of the cavity in a ligand may be able to override this criterion, *eg* the crown polyethers. The species 18-crown-6, shown in *Fig. 18a*, has a hole diameter of 2.6–3.2 Å (10 Å = 1 nm), and the stability constants of its metal complexes are shown in *Fig. 18b*.

The bases of this class are small, of high electronegativity, difficult to oxidise and of low polarisability.

FIG. 18. (*a*) The crown polyether, 18-crown-6. Hole diameter = 2.6 – 3.2 Å
(*b*) Stability of 18-crown-6 complexes.

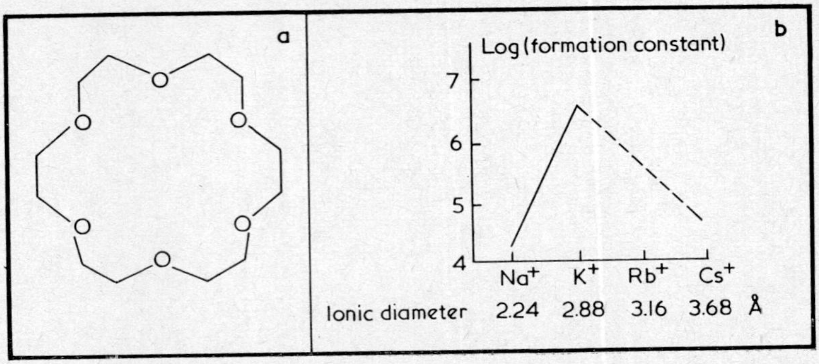

Class (*b*)—these metals are d° centres of low charge, *eg* Cu^+, Ag^+ or Au^+ or are transition metals with incomplete d shells, *eg* Cu^{2+} or Fe^{3+}. They prefer to bind to heavy halides, cyanide or sulphide rather than to oxygen donors. The bonds formed have a high covalent character and the bases are large, highly polarisable, of low electronegativity and are readily oxidised.

Since the original classifications were made, chemists have wished to categorise other acids and bases and so experimental methods have been devised for determining the 'hardness' or 'softness' of a species. These involve the use of a standard reaction. For base classification this may be

$$BH^+ + CH_3Hg^+ \rightleftharpoons CH_3HgB^+ + H^+$$

The hydrogen ion is 'hard' whereas the methylmercury ion is 'soft'. Thus, a 'hard' base will shift the above equilibrium to the left and a 'soft' base will shift it to the right.

Acids may be classified according to the order of stability of their halides, a 'hard' acid binding $F^- > Cl^- > Br^- > I^-$ whereas for a 'soft' acid the order is reversed.

Unfortunately, from the point of view of the chemist, the 'hardness' or 'softness' of a site does not depend only on the nature of the atom at the site but also on its environment. It has been shown that 'soft' bases tend to group together on a given central atom as do 'hard' bases. This mutual stabilising effect is known as *symbiosis*. Thus the addition of 'soft' substituents can 'soften' an otherwise 'hard' centre. For example, B^{3+} is borderline but BF_3 is 'hard' and BH_3 is 'soft'. This can be readily explained since the 'hard' F^- forms a complex which is largely ionic and so the charge on the B atom is high and, as we have pointed out, this leads to 'hardness'. 'Soft' H^-, on the other hand, donates charge extensively to the B atom by covalency and so decreases its charge and increases its 'softness'.

Table 9 shows some of the metal–ligand bonds that are known in biological systems.

A much more mathematical treatment of HSAB in terms of long and short range forces has been applied by Schwarzenbach to make quantitative predictions about the bonding between different groups.

Other effects seen in complex formation *in vitro* can also be applied to the *in vivo* situation.

- Stereochemical effects.
  A molecule cannot bind to a metal atom unless it can bend in such a way that the coordinating atoms can come close to the normal bonding distances from the central atom.
- The size of the ring formed.

Table 9. Some metal–ligand bonds known to be formed in biological systems.

Metal	Ligand	System
Mn	Imidazole	Pyruvate decarboxylase
Fe	Porphyrin, imidazole	Haem, oxidases, catalases
	Sulphur ligands	Ferredoxins
Co	Corrin ring	B_{12} enzymes
Cu	$\rangle N^-$ group	Haemocyanin
Zn	$-NH_2$, imidazole	Insulin, carbonic anhydrase
	$(-RS^-)_2$	Alcohol dehydrogenase
Pb	$-SH$	Aminolevulinic acid dehydratase

Rings of five or six atoms are the most stable and so most common.

- The chelate effect.

It has been known to chemists for a long time that the complexes of ethylenediamine, $NH_2CH_2CH_2NH_2$, were easier to prepare than those of ammonia and were less easily decomposed by acid or heat. Also, although benzaldehyde and phenol are poor ligands, salicylaldehyde (*o*-hydroxybenzaldehyde) is a very good chelating ligand (*see Fig. 19* for structures).

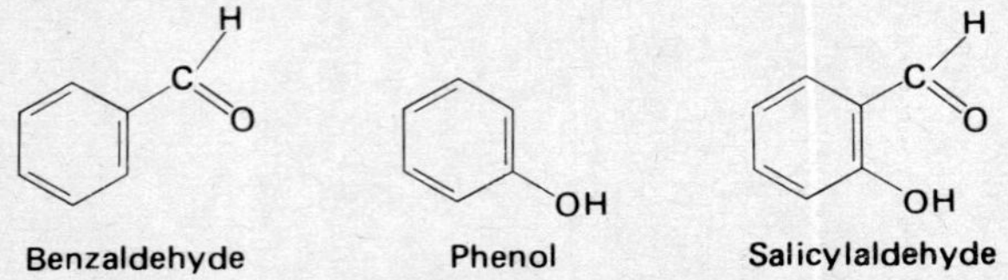

FIG. 19. The structural formulae of benzaldehyde, phenol and salicylaldehyde.

The additional stability of a chelate complex is a statistical effect which may be looked at from two points of view.

(*i*). Consider the reactions

$$M(H_2O)_n^{q+} + 2NH_3 \rightarrow M(NH_3)_2(H_2O)_{n-2}^{q+} + 2H_2O \quad 1$$

$$M(H_2O)_n^{q+} + NH_2CH_2CH_2NH_2 \rightarrow M(NH_2CH_2CH_2NH_2)(H_2O)_{n-2}^{q+} + 2H_2O \quad 2$$

In reaction (1) the change in the number of molecules is zero whereas in (2) the change is +1 molecule. Thus the randomness of the system is increased by reaction (2) and so it is favoured.

(*ii*). The second molecule of NH_3 to be bound to the metal has to be picked from the bulk of the solution whereas the second end of the ethylenediamine molecule is held near the metal and so its binding is more probable.

The above chemical principles have been used to make a great many predictions about bio-systems, *eg* of bond strengths and heats of reaction or of sites of binding

- CN^- binds 'soft' cations through carbon and 'hard' cations through nitrogen.
- 'Soft' metals are chosen for implantation in joint repairs, *etc*, since their ions form only weak bonds to the 'hard' ligand H_2O and so there is little tendency for the parent metal to dissolve.
- 'Soft' base poisons such as CN^-, CO, AsR_3 or S^{2-} hold metals in their lower oxidation states, to which they bind preferentially, and so prevent their function in redox enzymes.

The living system, however, is immensely complicated and does not always act in the way we might expect. We know that, in aqueous solution, zinc(II) binds the halides in the order $F^- > Cl^- > Br^- > I^-$ but in carbonic anhydrase, a metalloenzyme that controls the reactions $CO_2 + H_2O \rightleftharpoons HCO_3^- + H^+$ or $CO_2 + OH^- \rightleftharpoons HCO_3^-$, the binding order is reversed. We might at first attribute this to the 'softening' of the Zn^{2+} by the enzyme environment. However, CN^-, which is very 'soft' is bound no more strongly to the enzyme zinc than to the free Zn^{2+} and also NO_3^-, CNO^- and N_3^-, none of which is particularly 'soft', are bound exceptionally strongly to the enzyme zinc. The above ions are similar to the reactants and products of the metalloenzyme catalysed reaction, CO_2, CO_3^{2-} and HCO_3^-, and the enzyme seems to be tailored to accommodate these shapes of ions in a pocket of about 4.5 Å in length alongside the zinc ion.

The study of complex formation reactions

Complex formation can be described in terms of formation constants* which can be readily measured potentiometrically. From these the free energy of the reaction can be calculated since $\Delta G^\ominus = -RT \ln K$. The heat change of a reaction may be measured calorimetrically and so $\Delta H^\ominus$ calculated. Thus the remaining thermodynamic parameter, $\Delta S^\ominus$, the entropy change, can be found since $\Delta S^\ominus = (\Delta H^\ominus - \Delta G^\ominus)/T$. Strong complex formation is expressed by a high value of formation constant and is favoured by a high negative value of $\Delta H^\ominus$ and/or a high positive value of $\Delta S^\ominus$.

All processes be they chemical or biological are accompanied by heat changes many of which may be measured calorimetrically. The application of calorimetry to living systems is relatively new but

* In this publication the terms 'stability constant' and 'formation constant' are used synonymously to mean, for the reaction $A + B \rightleftharpoons AB$, $K = [AB]/[A][B]$.

already it can provide useful information. Whole organism studies on bacteria yield the following facts: (*i*) microbial degradation always involves the evolution of heat and so (*ii*) when bacterial growth is curtailed, *eg* by the exclusion of nutrients or insertion of antibiotics, there is a decrease in heat output; (*iii*) the heat characteristics of a particular type of bacterium are unique and reproducible and so can be used as a 'fingerprint' to identify the organism.

Calorimetry has also been applied to higher organisms where, although the heat per gram is less than for bacteria, the thermograms are still specific and depend on external agents, age *etc*. Bio-inorganic thermo-chemistry can be used to indicate antibody activity, for drug screening, for membrane studies or for kinetic studies of metalloenzymes, for example.

Studies of the rates of biochemical reactions may also yield important information about their mechanisms. Two features distinguish biochemical reactions from conventional organic reactions, *ie* their high velocity and their high selectivity. *In vitro* inert complexes, mainly those of Cr(III) and Co(III), take part in reactions with half-lives of more than about 60 s under normal conditions whereas the reactions of labile complexes are over in much less than 60 s. *In vivo* reactions usually have half-lives of less than 1 s and may have first order rate constants of more than 10^{10} s^{-1}. Kinetic studies may give the order of reaction which gives a guide to the molecularity and the rate variation with temperature leads to a value of $\Delta H^{\ominus}_{\text{activation}}$ which yields information about the transition state complex.

More sophisticated complexes *in vivo*

Complexes in solution

Most *in vivo* metal ions are bound to proteins but, of course, may also bind to amino acids, peptides, nucleic acids, added drugs or contaminating ligands. As discussed previously many ligands are likely to compete for each metal ion and the complex which is formed depends on the properties of both the metal and its environment.

Ligands exist which possess several groups which may have selectivities for different types of metal ions. For example, a molecule containing both a sulphydryl group and a carboxylate group may form a bond at one end involving the former donor and a 'soft' metal and also another bond in which the latter donor is complexed to a 'hard' metal so giving a mixed-metal complex. Human serum albumin is such a species having different sites for binding copper and zinc to each molecule.

At present more mixed-ligand (*ie* ternary) than mixed-metal complexes are known. In the definition of a mixed-ligand complex, water is not considered as a ligand but two other different ligands must be attached to the same cation. Mixed chelation occurs commonly in biological fluids, the stabilities of the species being determined by several factors:

(*i*). A study of the statistics of complex formation shows that mixed-ligand complexes are favoured.

The complex MA_4 can only be formed in one way whereas, assuming four different ligands, A, B, C and D binding to M with equal bond strengths, MABCD can be formed in $4 \times 3 \times 2 \times 1$ $(= 4!) = 24$ ways and so is more likely to be produced.

However, in general, ternary complexes are even more stable than the statistics would predict.

(*ii*). Stability is increased by charge neutralisation, *eg* $[\text{Zn cys·his}]^-$ is more stable than $[\text{Zn cys}_2]^{2-}$ (cys = cysteinate and his = histidinate anions of amino acids).

(*iii*). If A is a large ligand, the formation of MA_2 may be prevented by steric hindrance but if a second smaller ligand is available MAB may form.

(*iv*). Electronic effects such as the formation of π-bonds may be helpful to mixed-ligand complex formation.

(*v*). We have already discussed the *symbiotic effect* which means that two different ligands will only bind to a metal if they are of a similar degree of 'hardness'.

(*vi*). Also both ligands must prefer the same geometrical configuration of the metal ion, *eg* if one ligand prefers square planar and one octahedral then the ternary complex will probably not exist.

The techniques applied to the study of mixed-ligand complexes are essentially the same physicochemical methods as are applied to simple ligand complexes. Most of the quantitative information obtained so far has been from precise potentiometry. This yields the formation constant of the species $M_pH_qA_rB_s$ formed according to

$$pM + qH + rA + sB \rightleftharpoons M_pH_qA_rB_s$$

The concentration stability constant, β_{pqrs}, is measured at a constant temperature and constant background ionic strength,

$$\beta_{pqrs} = \frac{[M_pH_qA_rB_s]}{[M]^p[H]^q[A]^r[B]^s}$$

and may be used to suggest whether a reaction will occur. For example, consider

$$M(\text{L–A}) + (\text{L–A}) \rightleftharpoons M(\text{L–A})_2 \qquad 1$$
$$M(\text{L–A}) + (\text{D–A}) \rightleftharpoons M(\text{L–A})(\text{D–A}) \qquad 2$$

4

where (L-A) and (D-A) represent the two optical isomers of the ligand A. If β for reaction (2) is greater than for reaction (1) then the mixed-ligand complex is formed. An example of this is the case of cobalt–histidinate complexes in which Co(L-His)(D-His) is formed in preference to $Co(L\text{-}His)_2$ or $Co(D\text{-}His)_2$.

Because of their additional stability, mixed-ligand complexes are used a great deal by living systems, the enzyme-M^{n+}-substrate complexes which may activate the metal ion or bind the ligand ready for a selective reaction upon it being especially important. In a great many biological reactions one ligand attached to a metal is replaced by another involving the intermediacy of a ternary complex. The involvement of ternary complexes in biological processes ranging from active transport of ions across membranes to the uptake of nutrients by plants from soil has been shown and some of these will be discussed in more detail in the following chapter.

The *in vitro* study of ternary complexes has served as a model for, and led to a better understanding of, the function of metalloenzymes in biological catalysis.

Ternary complex formation involving copper is probably the system that has been most widely studied *in vitro* and a considerable amount has also been learnt about its mixed-ligand complex formation *in vivo*.

The normal adult content of copper is 50–120 mg but only in the high acidity of the stomach does free copper exist because of the successful competition of the proton for the ligands' binding sites. Human plasma contains about 15 μmol dm^{-3} copper. It has been observed that when all the amino acids present at physiological concentrations are added to a coppper solution a much larger proportion of ion is bound than would be expected from the individual binding capacities of these amino acids. Thus ternary complexes must be formed, the most probable ones involving histidine, threonine, cystine, glutamine and asparagine. A binding site which is very selective for copper is found in serum albumin which is the immediate transport form for this metal. Serum albumin, a chain of about 600 amino acids, has been extensively studied and the copper binding site is now known to be as shown in *Fig. 20*.

In 1912 Kinnear Wilson published findings on the inherited disease which has been named after him. Wilson's disease occurs in about 1 in 4 000 000 people in the US and is due to a breakdown of copper metabolism causing deposition of the metal in the tissues including the liver and brain. The symptoms may resemble those of acute liver disease or of a brain disturbance causing a shaking and lack of control of the limbs.

The exact nature of the genetic defect causing this disease is not known but the symptoms can be alleviated by removal of the excess

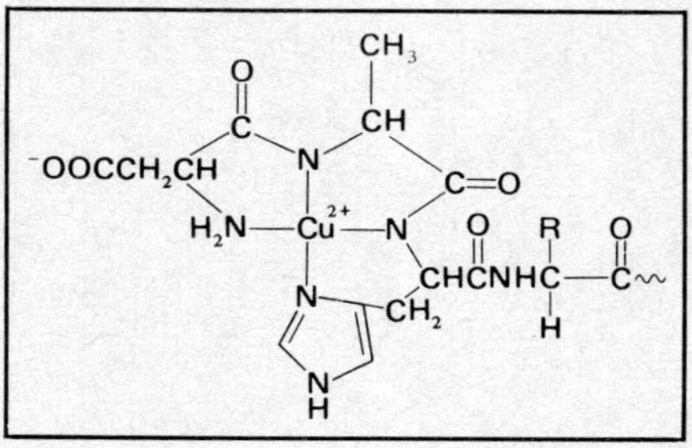

FIG. 20. The copper(II) binding site of human serum albumin

copper. This removal can be achieved by use of several chelating agents, *see* Table 10, the most effective being D-penicillamine. Unfortunately, with all of these drugs, unpleasant side effects occur and, since these patients will require treatment for the rest of their lifetimes, the situation is hardly satisfactory. Also, because of the lack of selectivity of the drugs, other biometals will be removed and have to be carefully supplemented.

Table 10. Chelating agents which may be used for the removal of excess *in vivo* copper.

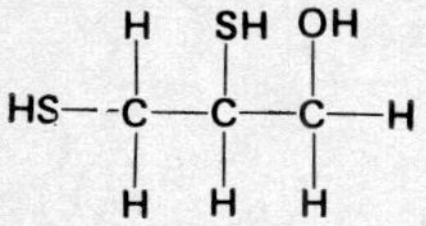

BAL (2,3 dimercaptopropanol)

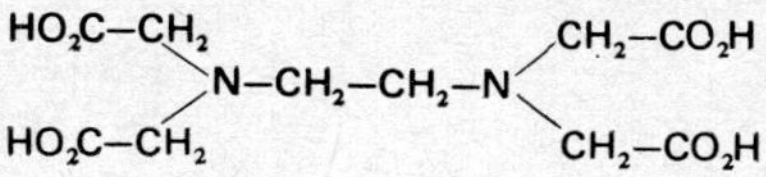

EDTA (ethylenediaminetetraaceticacid)

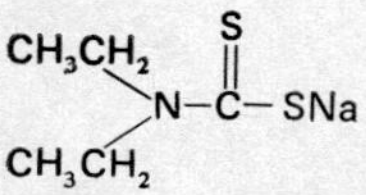

sodium diethyldithiocarbamate

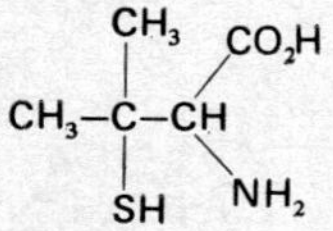

D-penicillamine (reduced form)

$H_2N{-}CH_2{-}CH_2{-}NH{-}CH_2{-}CH_2{-}NH{-}CH_2{-}CH_2{-}NH_2$

triethylenetetramine

However, B. Sarkar and his coworkers in Canada recognised that if they could duplicate the copper selective site of human serum albumin they would have a much superior drug for the treatment of Wilson's disease, since it ought not to remove a significant amount of other essential metals such as iron or zinc. Serum albumin itself cannot be used because the large molecule is not able to cross biological membranes to be excreted. Thus Sarkar and coworkers have prepared several small peptides which have similar chelating possibilities to the binding sites in serum albumin. An assessment of the potential effectiveness of these ligands is given later (*see* p 57).

Metalloenzyme complexes

Simple metal ions alone will catalyse a large number of organic reactions in solution but metalloenzymes are much more efficient and selective to one reaction. Normally an enzyme reaction will be about 10^9 times faster than an uncatalysed reaction.

Enzymes, of course, are proteins and so the normal difficulties associated with protein investigation are attendant upon their study. For example, a slight change in temperature or pH can greatly alter the structure of a protein and so the normal metal binding site may no longer exist. The identification of the binding groups is always difficult since many potential ligand groups are present in different natural amino acids, *eg* amino nitrogens, imidazole nitrogens, carboxylate groups, alcohol groups or sulphydryl groups. However, the presence of a transition metal makes the study easier because its characteristic properties provide very useful information concerning the nature and symmetry of the metal binding site.

Because of the large variety of ligand groups available, metals in metalloenzymes are generally bound to several different groups so allowing the use of ternary complexes as models. Even the determination of a stability constant for a metalloenzyme–metal interaction is made very difficult by the change in pH causing a conformational change and so a change in binding sites.

However, mechanisms have been proposed for several enzymic reactions. The enzyme carbonic anhydrase, which contains Zn^{2+}, catalyses a reaction of great physiological importance, the hydration of carbon dioxide or the reverse reaction, the dehydration of carbonate

$$\begin{array}{l} CO_2 + H_2O \rightleftharpoons HCO_3^- + H^+ \\ \qquad\qquad\qquad\quad \upharpoonleft\downharpoonright \\ \qquad\qquad\qquad\quad CO_3^{2-} + H^+ \end{array}$$

The turnover number (under saturation conditions) for CO_2 hydration is about 10^6 per second, the uncatalysed hydration rate being 7.0×10^{-4} per second. The molecular weight of the enzyme is about 30 000 and its amino acid composition and three dimensional crystal structure have been determined. The Zn^{2+} would appear to lie near the centre of the molecule in a cleft where it can have three imidazole nitrogen ligands and have the fourth tetrahedral site open to the surrounding medium. The proposed mechanism for the enzyme reaction is shown in *Fig. 21*.

The active site of an enzyme consists of the polar groups necessary for its activity and is generally situated at the end of a corridor of low

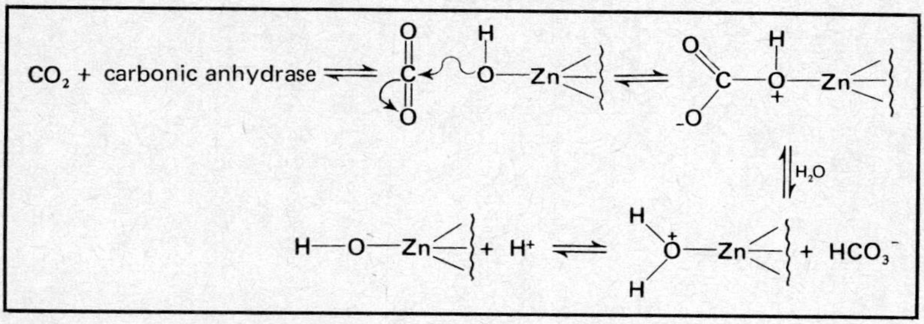

FIG. 21. The proposed mechanism for the catalytic action of carbonic anhydrase.

dielectric constant into which the substrate can fit. The active site itself is believed to be in a geometry approaching that of the transition state of the particular reaction being catalysed. This is called the 'entatic state'. Thus the enzyme can provide a new reaction pathway with a lower activation energy and so greatly speed up the reaction. Very frequently, organic reducing agents have the correct redox potentials for a biological reaction to occur, *ie* the reaction is thermodynamically feasible, but due to kinetic factors, *ie* a high activation energy, the reaction does not occur. This is precisely the problem which an enzyme is designed to overcome (*see Fig. 22*).

As can be seen the catalyst lowers the activation energy of both

FIG. 22. The effect of a catalyst on a chemical reaction.

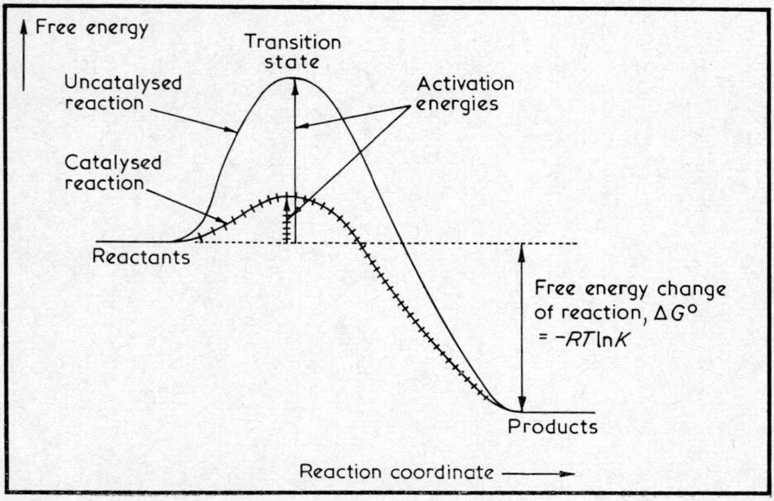

the forwards and backwards reaction and so although it speeds the reaction up it cannot change the position of equilibrium.

Iron metalloenzymes

The chemistry of an element is fixed by the nature of its atoms but its biology may change within chemical limits as the evolutionary process advances. The *in vivo* function of iron is extremely complicated and, of course, very important.

In solution Fe(III) and Fe(II) are the only important oxidation states although all the states from I to VI are known and even −II is found in species such as $Fe(CO)_4^{2-}$. The most common ligand arrangement with iron is octahedral six coordinate but tetrahedral four coordinate is also known. In biomolecules, however, the environment has a pronounced effect on the chemistry of the metal even to the extent of determining its oxidation potential Iron can be fixed as the more highly charged Fe^{3+} by surrounding it by a 'hard' environment such as water. Thus Fe(II) solutions readily hydrolyse to the more stable Fe(III)(aq) complex.

The serum iron transport protein, transferrin, carries iron to the bone marrow where it also catalyses the insertion of the iron into a porphyrin ring to give a haem group. Transferrin binds Fe(III) 10^{27} times more strongly than Fe(II) and so the Fe(II)–transferrin complex is very readily oxidised to Fe(III)–transferrin. Formation constants have been obtained from pH studies on transferrin, which has a molecular weight of about 77 000 and binds two iron atoms per molecule. The stability constants suggest that the binding groups are phenoxy groups of tyrosyl residues, and imidazole groups of histidyl residues. In fact, it is believed that each of the two iron atoms in the complex is bound to three tyrosyl and two histidyl residues.

Human haemoglobin, the oxygen carrying species, has a molecular weight of about 65 000 and the protein part is made up of two α and two β units each of which contains one polypeptide chain, the amino acid sequence of which has been determined, and one protoheme IX group as in *Fig. 12a.* The haem group is located in a hydrophobic pocket in the protein structure. As was discussed in Chapter 3, the fifth axial position is supplied by the imidazole group of a histidyl residue and the sixth position may be filled by O_2 (or even CO or CN^- in cases of poisoning). Thus carbon monoxide and cyanide are extremely potent enzyme poisons excluding oxygen from the haemoglobin.

There are many different haemoglobins having only slight variations in structure. For example, Nature has made sure that adult and foetal blood have different haemoglobins which have their maximum uptakes of oxygen at different pressures so facilitating transfer of oxygen from mother to foetus.

Co-enzymes and their complexes

Most enzymes require cofactors for their activity which may be either a simple metal ion or a complex organic molecule called a co-enzyme. Many co-enzymes are vitamins, or their derivatives, and generally act as intermediary carriers of electrons, or of specific functional groups that are transferred in the enzymatically catalysed reaction.

An extremely important reaction in amino acid metabolism is transamination—the transfer of an amino group from an α amino acid to an α keto acid.

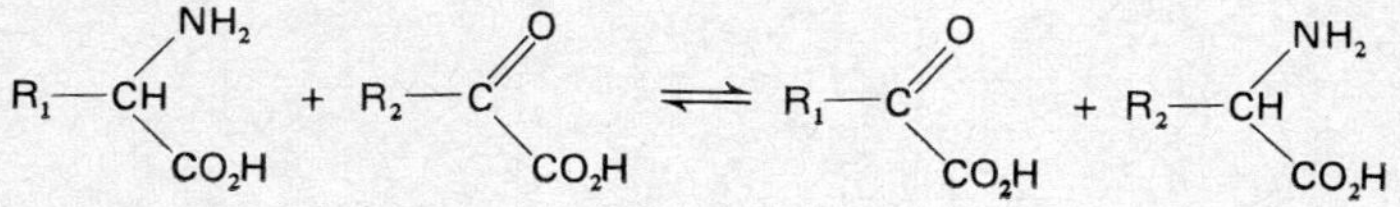

This reaction is catalysed by a transaminase which requires the co-enzyme pyridoxal phosphate, a derivative of vitamin B_6, *see Fig. 23.* The aldehyde group allows pyridoxal phosphate to act as a carrier of amino groups in a reaction as follows: (Ⓡ represents the transaminase pyridoxal phosphate complex)

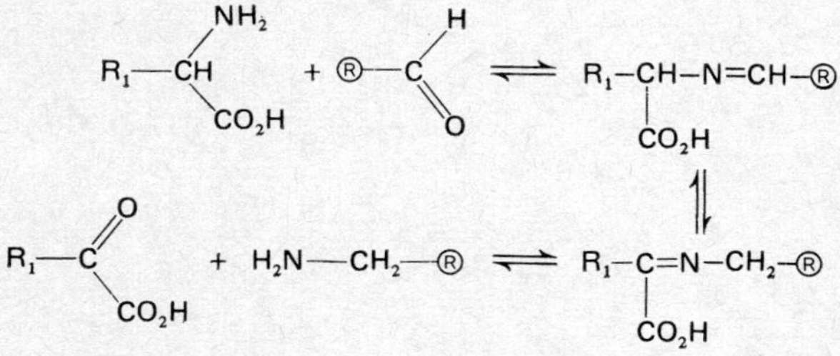

A derivative of vitamin B_{12}, 5-deoxyadenosyl vitamin B_{12}, on the other hand, catalyses reactions of the type

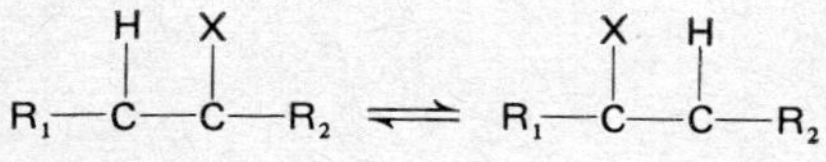

In the mechanism it is thought that the Co(II) of the co-enzyme may be reduced to Co(I) which can act as a carrier of alkyl groups from one atom to another. Co(I), of course, would be very unstable in normal aqueous chemistry.

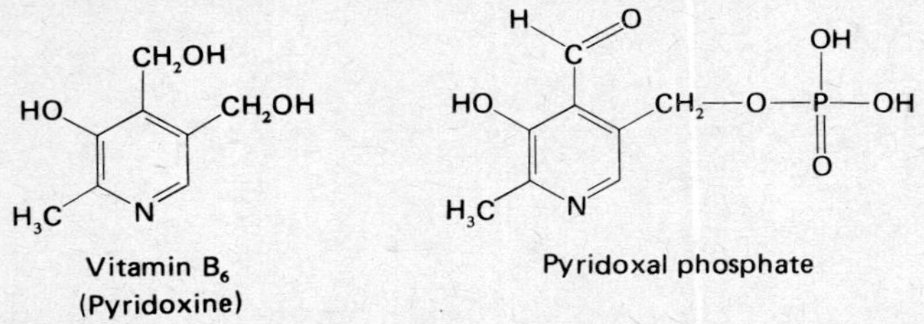

FIG. 23. Structure of vitamin B_6 and its active derivative.

Choice of chelating agent for removing an excess of a metal ion

People living in a modern industrial society may be exposed to excessive concentrations of both non-essential and essential metal ions as well as the non-metal pollutants such as sulphur dioxide. Those causing most concern are lead, mercury and cadmium but cobalt, iron, copper, zinc and manganese are also encountered in quantities sufficient to cause toxic reactions.

As well as excesses, deficiencies of essential metal ions are known, iron being the most common case. Thus methods are required for both removing metals from, and administering metals to, the body.

The biological effect of a toxic metal ion depends not only on the nature of the ion but also on its rate of absorption, distribution, deposition and excretion. Most of the harmful effects are caused by the binding of the toxic ion to a metabolically important group often with the displacement of an essential trace metal ion. Soft metals such as mercury or arsenic will prefer to bind to ligands in the order $S > N > O$ whereas hard metals such as calcium or magnesium will bind $O > N > S$ and those such as zinc or lead will be intermediate in character. The damage caused may not be reversible even when all the excess toxic ion is removed, *eg* the brain damage caused in children by excessive lead uptake.

The requirements for a good chelating agent for the removal of a toxic excess of a metal ion may be summarised as follows:

(*a*). The drug must not be destroyed by the method of administration, the method of choice being oral. Peptides, for example, could not be given by this route since they would be digested in the stomach.

(*b*). It must be as selective as possible otherwise, under prolonged treatment, other metals may have to be supplemented.

(*c*). It must be small enough to be able to pass through membranes to reach the metal at its site of binding.

(*d*). It must be able to remove the metal from its biological binding site by binding more strongly to it.

(*e*). Both the chelating agent and its complex to the toxic metal must be soluble and non-toxic. Zinc or cadmium complexes with the drug BAL are more poisonous than the free metal ions!

(*f*). The drug must not transport the metal ion to a place where it may do more damage. The treatment of lead poisoning with BAL can cause lead to be carried across the blood/brain barrier to produce brain damage.

Some of the chelating agents which are used to treat metal excesses are shown in Table 11.

Table 11. Some chelating agents used to treat metal excesses.

Ca	EDTA (ethylenediaminetetraaceticacid)
Fe	Desferrioxamine or Na_2[Ca EDTA]
Cu	D-penicillamine or Na_2[Ca EDTA]
Co, Zn	Na_2[Ca EDTA]
Cd, As	BAL(dimercaptopropanol)
Hg	BAL or *N*-acetylpenicillamine
Pb	D-penicillamine or Na_2[Ca EDTA]
Be	Aurintricarboxylic acid
Tl	Diphenylthiocarbazone
Ni	Sodium diethyldithiocarbamate
U	Na_2[Ca EDTA]
Pu	Na_3[Ca DTPA] (Na_3Ca diethylenetriaminepentaacetate)

BAL—British Anti Lewisite—was developed as a protection against the poison gas Lewisite, Cl–CH=CH–$AsCl_2$, which acted by binding to the sulphydryl groups of some enzymes with resulting inactivation. However, BAL binds more strongly to the arsenic than does the enzyme and the BAL complex can be excreted.

The factors underlying the design of a suitable drug for the removal of a particular metal are fairly well understood. Thus, if sufficient data are available on the interactions of the metal ion with all the biological ligands, it should be possible to use chemical theory to design an effective therapeutical. Even if the relevant stability constants are not available, selection of a drug may be made by considering the metal ion from an HSAB point of view while also remembering that the stereochemistries of the ligand and the metal must be compatible.

Unfortunately, because of the great complexity of the life process, a high stability constant is not sufficient guarantee that a chelating agent will be effective under biological conditions. For example, $HgEDTA^{2-}$ has a higher formation constant than $ZnEDTA^{2-}$ but EDTA removes zinc not mercury.

Computer models in medical research

If we have sufficient concentration and formation constant data, including those for ternary complexes, it is possible to use computer programs to do the extremely complicated, and tedious, calculations necessary to build up a model of a biological system.

The general definition of a model is anything which simulates a real system and attempts to imitate reality. Pharmaceutical models are often mathematical. The scientific philosophy requires that hypotheses, drawn up in the light of observation, be used to make predictions which are then confirmed or refuted by more experiments. Thus models facilitate the design of new experiments that test both theories and hypotheses in depth.

There are sound reasons for applying computer models to medical research. During the last 15 years the costs of discovering and bringing to market a new drug have trebled so that each new product costs many millions of pounds. As distinct to the blanket screening of all chemicals against all diseases, companies are now devoting their attentions to new approaches that help to select 'lead' molecules for further research and computers are being extensively used in this area.

In principle, one wishes to plot biological activity as a function of some physical or chemical parameter (for example, activity in antiviral screens *versus* solubility in chloroform—remember that cell membranes are often likened to an organic solvent). There are two basic approaches for assigning a value to this chemical or physical parameter. The Hansch approach involves structure activity relationships which assumes that the lipophilicity (*ie* membrane solubility) of a compound is the sum of the contributions from its many component groups as calculated using a computer. An alternative approach used in this monograph is that of concentration activity relationships in which the concentration of each species present in each phase (for example, cell membrane or aqueous fluid bathing the cell) is calculated from lists of the total concentrations of all metals and ligands present and all their equilibrium constants (for solution complexes, solvent distributions and partial pressure constants when gases are involved). Currently such models are capable of handling more than 9000 species in a simulation of blood plasma. Although nature is never truly at equilibrium, in order to achieve high efficiencies of energy conversion, most biological systems operate near to reversible equilibrium conditions and so the use of equilibrium constants is justified.

The aim of such models is to plot hill-shaped activity *versus* concentration curves whereby the optimum system is at the summit of the hill. Adverse biological properties (for example, toxicity) may be imagined as a second hill near to the first. Judicious use of computer models can not only optimise the route to the correct activity summit but also minimise the toxicity hill involvement.

It must be noted that the model is not creating new chemical facts but rather reflecting already known facts in a more economic, effective and enlightening manner. Thus the choice of chelating agent for removing a metal ion (*see* previous section) can be quantified, the dose optimised and even the amount of supplementation of other metals calculated. Promising ligands from these theoretical studies are then put forward to animal and clinical trials.

An example—the therapy of Wilson's disease

Previous sections of this chapter have described the symptoms arising from the deposition of copper in Wilson's disease, agents to remove this copper, ligands which mimic the binding site for copper(II) in human serum albumin, general factors to be considered in choosing an agent to remove a metal ion and the uses of computers in medical research. All of these aspects can be illustrated by using computer simulation to consider this copper excess condition and its therapy in closer detail.

Plasma contains copper ions complexed in four forms:

(*i*). Those firmly complexed in metalloproteins, *eg* ceruloplasmin (the copper ion—donor group bonds being inert).

(*ii*). Those more loosely complexed to other proteins, *eg* serum albumin.

(*iii*). Those associated with the many low molecular weight ligands present.

(*iv*). Those existing as aquated ions.

The latter three forms may be considered as being a series of labile complexes which are in competitive chemical equilibrium.

Excretion involves the labile low molecular weight complexes, (*iii*).

The first step towards copper removal is the liberation of copper deposits such as the copper firmly and inertly complexed to ceruloplasmin or the use of powerful ligands to remove the copper(II) from serum albumin and to form more low molecular weight complexes. If these complexes are small and preferably charged then they can be excreted through the kidneys.

The copper mobilising influence of the agents listed in Table 10 can best be assessed by a detailed analysis of all the metal–ligand equilibria occurring in the plasma of a patient to whom the drug has been administered. However, experimental measurements *in vivo* are not possible because the complex concentrations are orders of magnitude below those amenable to even the most sophisticated analytical equipment and because there are many thousand species present. Thus simulation of these equilibria using a knowledge of each individual equilibrium system (determined at higher concentrations *in vitro*) and high speed computers is the only means of estimating these equilibrium concentrations *in vivo*.

Table 12. The percentage distribution of the metal ions Ca^{2+}, Cu^{2+}, Fe^{3+}, Pb^{2+}, Mg^{2+}, Mn^{2+} and Zn^{2+} amongst low molecular weight ligands in human blood plasma as found by computer simulation.

Complex	Percentage of the total metal in the low molecular weight fraction
Ca.carbonate.H^{+}	9
Ca.citrate^{-}	4
Ca.lactate^{+}	3
Ca.phosphate^{-}	3
Ca.carbonate	2
Cu.histidinate.cystinate^{-}	21
Cu.histidinate.cystinate.H	17
Cu.histidinate$_2$	11
Cu.histidinate.threoninate	8
Cu.histidinate.valinate	5
Cu.histidinate.lysinate.H^{+}	5
Cu.histidinate.alanate	4
Cu.histidinate.serinate	4
Cu.histidinate.phenylalanate	3
Cu.histidinate.glycinate	3
Cu.histidinate.leucinate	2
Cu.histidinate.glutamate^{-}	2
Cu.histidinate.glutaminate	2
Cu.histidinate.ornithinate.H^{+}	2
Cu.histidinate.prolinate	1
Cu.histidinate.isoleucinate	1
Cu.histidinate.tryptophanate	1
Fe.citrate.OH^{-}	99
Fe.citrate.salicylate^{2-}	<1
Fe.citrate.glutamate^{2-}	<1
Fe.citrate.oxalate^{2-}	<1
Pb.cysteinate	80
Pb.cysteinate.citrate^{3-}	7
Pb.cystinate.H^{+}	5
Pb.cysteinate.phosphate.H^{2-}	3
Pb.carbonate.H^{+}	2
Pb.cysteinate$_2H^{+}$	2
Pb.carbonate	<1
Mg.carbonate.H^{+}	6
Mg.citrate^{-}	5
Mg.carbonate	2
Mg.lactate^{+}	2
Mg.phosphate.H	1
Mn.carbonate.H^{+}	24
Mn.citrate^{-}	10
Mn.carbonate	2
Mn.oxalate	2
Mn.phosphate.H	1
Zn.citrate, cysteinate^{3-}	43
Zn.cysteinate$_2^{2-}$	19
Zn.cysteinate.histidinate^{-}	12
Zn.cysteinate	3
Zn.histidinate^{+}	3
Zn.cysteinate$_2H^{-}$	1
Zn.histidinate$_2$	1
Zn.cysteinate.glutaminate^{-}	1

The major low molecular weight complexes of metal ions in the plasma of a healthy human are computed to be those shown in Table 12. In Table 13 we list the main complexes formed when

Table 13. The main complexes formed when chelating agents are administered to plasma.

Chelating agent	Metal ion	Complexes predominating
Trien	Cu(II)	$Cu.Trien^{2+}$
	Fe(III)	$Fe.Trien^{3+}$
EDTA	Fe(III)	$Fe.EDTA^{-}$, $Fe.EDTA.OH^{2-}$
	Pb(II)	$Pb.EDTA^{2-}$
	Mn(II)	$Mn.EDTA^{2-}$
	Zn(II)	$Zn.EDTA^{2-}$
Pen (reduced)	Pb(II)	$Pb.Pen^{0}$, $Pb.Pen.citrate^{3-}$
	Zn(II)	$Zn.Pen_2^{2-}$, $Zn.Pen.cysteinate^{3-}$
(oxidised)	Cu(II)	$Cu.Pen.histidinate.H^{0}$, $Cu.Pen.histidinate^{-}$.

Abbreviations used: Pen = D-penicillamine
Trien = triethylenetetramine
EDTA = ethylenediaminetetraacetic acid

D-penicillamine, triethylenetetramine, and for comparative purposes, the well known chelating drug ethylenediaminetetraacetic acid are administered. The new low molecular weight species occurring in Table 13 obtain the vast majority of their metal contents from proteins circulating in plasma (for example, serum albumin), and the remainder at the expense of the complexes listed in Table 12. Thus, the total low molecular weight content of metal in plasma increases when chelating drugs are administered.

The mobilising influence of the administered chelating agent is illustrated in *Fig. 24*. In order to avoid the problem of considering different patients having different plasma drug levels a mobilising index (PMI) can be defined as the factor by which, for a given concentration of chelating agent, the total low molecular weight fraction of metal ion is increased. It provides a convenient measure of the ability of a chelating agent to mobilise the metal from the labile protein bound fraction. PMI is computed as (the total low molecular weight species in the presence of drug)/(total low molecular weight species in normal plasma). This ratio effectively eliminates uncertainties arising from metal protein interactions that would be associated with absolute concentrations. Although all the functions have an upper limit imposed by the total exchangeable metal on the proteins, zinc is atypical in that this limit is rapidly approached and so it is likely to be important. This is because the low molecular weight complexes of zinc in plasma have a total concentration relatively

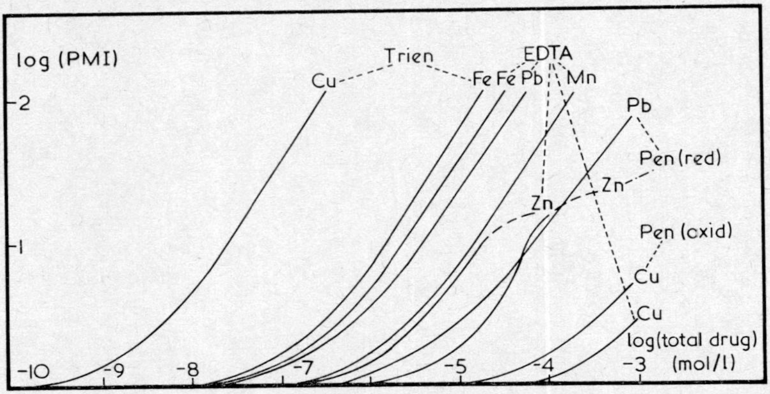

FIG. 24. The metal mobilising influence of administering Trien, Pen and EDTA to normal plasma.

close to the total protein reserve and so the curves for this metal ion are expected to level off (*see Fig. 24*).

The figure shows that Trien medication as low as 10^{-9} M begins to mobilise copper. The agent quoted for reference, EDTA, avidly binds iron(II), and manganese(II) but does not significantly influence copper and so is not a satisfactory therapeutical for the treatment of Wilson's disease.

Pen, in its reduced form, increases low molecular weight zinc levels at 10^{-6} M and also mobilises any lead present. Excretion of both these metals is promoted by this chelating agent; Pen is, in fact, the treatment of choice for lead poisoning. On the other hand, oxidised Pen only feebly binds copper at 10^{-5} M Pen and above, *ie* the copper mobilising influence of Pen is 10^{-4} to 10^{-5} that of Trien.

Thus Trien, the agent used as a substitute for Pen in treating Wilson's disease patients undoubtedly exerts its influence *via* a mobilising effect upon the labile copper pool in blood plasma.

At least two modes of action for Pen on copper metabolism are consistent with these model studies. Possibly the drug may activate a transport mechanism for concentrating copper from plasma. This is not excluded by the fact that renal rather than hepatic excretion is enhanced by Pen therapy although it does seem somewhat incongruous. It is far more likely that Pen releases copper from inert metalloprotein such as ceruloplasmin by reduction to copper(I). This copper would not otherwise be available to the labile equilibrium even in the presence of powerful chelating agents such as Trien. However, once the copper(I) becomes oxidised to copper(II) in plasma it enlarges the pool which is potentially capable of diffusion across the membrane. Experiments have shown (J. M. Walshe) that Pen is indeed capable of such a role when ceruloplasmin is present although some other source of copper may be invoked in Wilson's disease patients deficient in this metalloprotein. Walshe has also shown that Trien and Pen function in different body compartments and this is in complete agreement with these model studies.

It should be noted that the increased concentration of low molecular weight copper produced by these drugs is then available for excretion. However, the rate of removal of copper complexes from plasma and the route which they take may depend upon other factors such as their size and charge as well as upon the amount mobilised.

We can now comment on the mobilising/excretion potential of peptides such as those proposed on p 47. Computer models show that they do indeed mobilise copper better than Pen or EDTA although not as effectively as Trien (we might note that these ligands have the added advantage of being composed of naturally occurring components, amino acids) but, regrettably, form mainly neutral complexes (Trien forms $(\text{Trien Cu})^{2+}$) which are not conducive to renal excretion, but rather might encourage copper deposition in other parts of the body because their electroneutrality gives membrane penetration powers, thus allowing them to escape from the plasma–kidney stream.

Finally, we might note that computer produced simulations similar to those depicted in *Fig. 24* are useful in indicating likely bio-metal dependent side effects of administered chelating agents. For example, the figure suggests that Pen therapy of Wilson's disease will also modify plasma zinc concentrations.

This section may appear biased in favour of low molecular weight complexes (rather than metalloenzymes and labile protein–metal ion complexes) but this preference is intentional because it is only the low molecular weight complexes of metal ions which are absorbed from the intestine, which are transported between blood plasma and tissues, which are excreted in bile or urine and which are incorporated into metalloenzymes. (For example, in the syntheses of cuproproteins it is impossible to introduce Cu^{2+}(aq) ions from plasma at about 10^{-18} mol dm^{-3} whereas ternary mixed ligand complexes, which are a million times more concentrated than Cu^{2+}(aq), are built into the protein.)

5. Principles—Phenomenological

There are many medical conditions which are firmly established as being bio-metal dependent. However, it is often the case that the molecular chemistry prevailing (at the levels discussed in Chapter 4) is not known. Therefore, it is not surprising that we cannot begin to guess at a successful cure (as discussed in Chapter 6).

Nevertheless, the principles applying to one element are usually obeyed by others. These principles are outlined in this chapter (*a*) in the hope that keen observers can reveal many more examples, and (*b*) in the desire to attract bio-inorganic chemists to research the molecular chemistry of these observations so that this might well appear in Chapters 4 or 6 of a second edition.

Brown, in monograph number 17, has already described the 'unity and diversity' principle of modern bio-inorganic chemistry: there are many processes in microorganisms, plants and animals that use essentially the same chemistry but the diversity of colours, forms, and sizes are the result of chemical differences or variations based upon this unifying central theme. For example, both humans and plants require calcium and sulphur as essential elements and so have evolved mechanisms for ensuring the absorption and assimilation of these two. Yet, the biochemistries of humans living for long periods in an area with drinking water high in calcium have diversified and developed means of preventing its hyperabsorption (which would lead to atherosclerosis by calcium precipitation). So too, vines grown on the volcanic, highly sulphurous areas of Hungary, have diversified to bear grapes with a very low sulphur content (and thus produce Tokaj Furmint wines of minimal morning after hangover effects). There are occasionally Achilles' heels in this process; for example, although the human body can accommodate microbes that produce vitamin B_{12}, this essential compound cannot be used by humans unless it is passed out in stools and taken in again as a food impurity.

The origins of life have been described by Brown and discussed further in our Chapter 2. It is noteworthy that during this evolution, in spite of the 'organism rebelling against excesses' principle of the previous paragraph, our cells have evolved to be very similar indeed to the sea-water and sediments of the sea-bed from which our ancestors evolved (*see* Table 14): the same ions predominate in both the oceanic and living tissue systems (Na^+, K^+, Ca^{2+}, Mg^{2+} and Cl^-) and, furthermore, the concentrations are similar. Österberg has drawn attention to a useful comparison between extra- and intra-cellular compositions and sea-water and sediment respectively; the

Table 14. Some concentrations of ions in sea water, ocean sediments, blood plasma and intracellular fluid. Units: mmol dm^{-3}.

Ion	1 Sea water	2 Sediments	3 Plasma	4 Intracell	Ratios 1:2	3:4
Na^+	470	290	140	<10	2:1	14:1
Mg^{2+}	50	480	1	20	1:10	1:20
K^+	10	400	4	110	1:40	1:28
Cl^-	55	—	100	—	>100:1	>100:1
HPO_4^{2-}		20	1	100	1:100<	1:100

high K^+, Mg^{2+} and HPO_4^{2-} in the cell and in sediments (and *vice versa* for the extracellular and sea-water concentrations which instead have high Cl^- levels) are, he suggests, indicative of our cells having evolved from a crystalline phase—buffered mother liquor origin. Currently, there are many scientists who regard cell membranes as a liquid crystal and also there are several who regard the complete cell as a mixture of such liquid crystals; one is reminded of Yong's comment from the beginning of this century that 'a cell membrane is a boiling sea of fat possessing a fixed localisation'!

It appears that if, in some locality, an element becomes enriched more rapidly than cellular evolution can adapt to it, then, paradoxically, almost complete rejection occurs akin to that of the calcium and sulphur examples just described.

The cycling of elements in nature

Whilst there is considerable doubt whether organisms are reincarnated after death and reappear as another organism, at the *element* level the concept is certainly valid. Chapter 2 and *Fig. 3* have described the various types of elements in our bodies. These elements participate in a variety of cycles involving their uptake into man from his environment, as food, assimilation, uses *in vivo*, excretion and re-entry into the environment. *Figures 5* and *25* exemplify this pattern. In fact, each element will have several route modifications to the same cycle in respect of its existence in humans; these will depend on whether its participation is in the biochemistry of living man, of deceased man, of decaying cadaverous chemistry or of the cremated species.

Provided one defines a set of prevailing conditions, there will be a fairly reproducible biological half-life (the time it takes for half of the present atoms of an element to be removed—even if they are replaced by like atoms of the same element) for each element. Table 15 lists some common examples.

It is natural, but nevertheless naive, to suggest that for a body in a state of good health, one ought to be able to compile a table of 'normal' total concentration values for each element in each part of

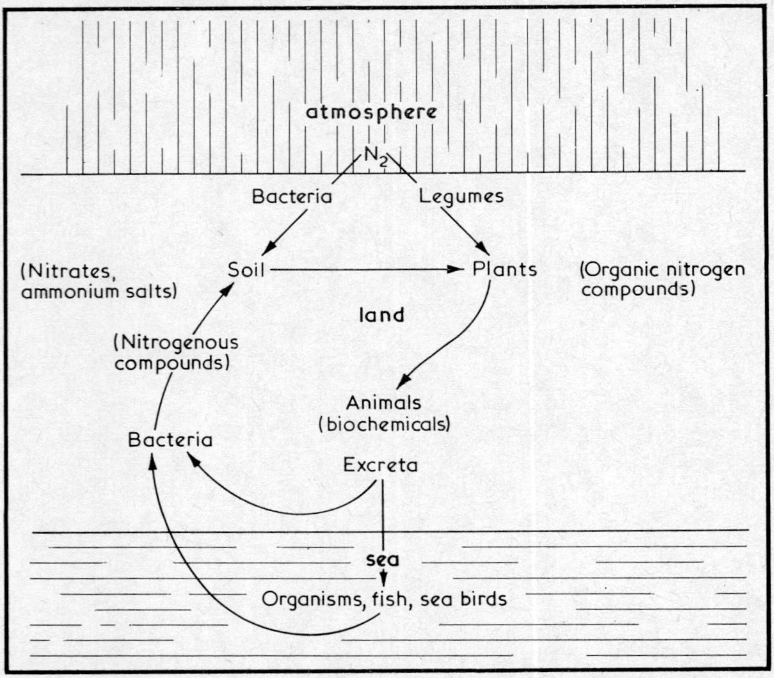

FIG. 25. All elements have a cycle in nature. This is an example of a cycle for a non-metal nitrogen (*see* also *Fig. 5* for the cycle for a metal). Note that the element can have different oxidation states in different parts of the cycle.

the body. In practice this is an exceedingly complex task. Let us assume that two healthy individuals at different ends of the country, have blood samples taken; in principle both ought to have 'normal' Na, Cu, Pb counts but the recorded figures may well vary because of their ages (we shall see later that many metals accumulate in the body with ageing), their current diseases and their past medical history, whether they are male, female, child or foetus, their familial and religious dietary trends, the local environment (air, water, food),

Table 15. The biological half-lives of some elements in man.

Element	Half-life	Element	Half-life
Hydrogen	19 d	Carbon	35 d
Sodium	16 d	Potassium	13 d
Calcium	5 y	Chromium	3 y
Manganese	3.5 d	Iron	66 d
Cobalt	8.5 d	Copper	68 d
Zinc	23 d	Iodine	112 d

the stimulation and/or antagonism of other elements (*see* later in this chapter), the time of day the sample was taken (later we shall describe the influence of circadian concentration changes), the means of withdrawing, bottling, and transporting the sample and eventually the accuracy of the analysis for the element concerned.* Under these circumstances, it is not surprising that the term 'normal value' is being dropped and replaced with 'reference value', this latter having the advantage that it does not suggest that an individual who provides a hyper or hypo figure is necessarily 'abnormal' but rather that further complementary tests are advisable (*eg* if the potassium analysis is lower than the reference value, then determine calcium and magnesium concentrations).

Inorganic nutrition

Eagerness to conquer the subtle chemical determinants of growth and vitamins necessary for health has often lead to a neglect of the inorganic requirements of the organism. Indeed, these organic growth factors usually mediate through inorganic complexes and so the kinetics of cellular growth and reproduction are dependent upon the limiting concentration of an element or vitamin in the diet. A noteworthy example is the dwarfism occurring in Eastern European countries; this is directly relatable to, and responds to treatment with, zinc in the diet.

Some of the principles of competitive complexing necessary for an element to become absorbed efficiently are understood. The actual mechanisms of assimilation are mainly unknown. Let us consider the cases of zinc supplementation (to encourage wound healing, to offset the zinc excreted by muscle catabolism during fevers *etc*) on an empty stomach and then again the zinc absorbed from a meal.

The actual site of absorption of transition metal ions is the small intestine (the duodenum and the jejunum). The cells lining the intestine may be likened to a layer of mixed organic solvent (lipids and proteins) and a metal ion must be complexed by ligands present in the intestinal juices to give a neutral (*ie* uncharged) metal complex capable of being dissolved in this organic layer. The aqueous layer bathing this lining is at a pH of about 6–7. Thus, it is becoming established practice to administer metal ions as their complexes on an empty stomach (*ie* between meals), the complexing ligand being chosen (*a*) such that it gives a high concentration of neutral complex in aqueous solution at pH 6–7, and (*b*) such that it does not complex

---

* This last stage might appear foolproof and reproducible but we must remember that, in practice, the analysis is done in the presence of dozens of other elements in the plasma which may influence the results. For example, the same sample of blood was recently circulated to pathology laboratories in the UK and the lead analyses returned varied from 9 to 130 μg per 100 cm^3 of lead!

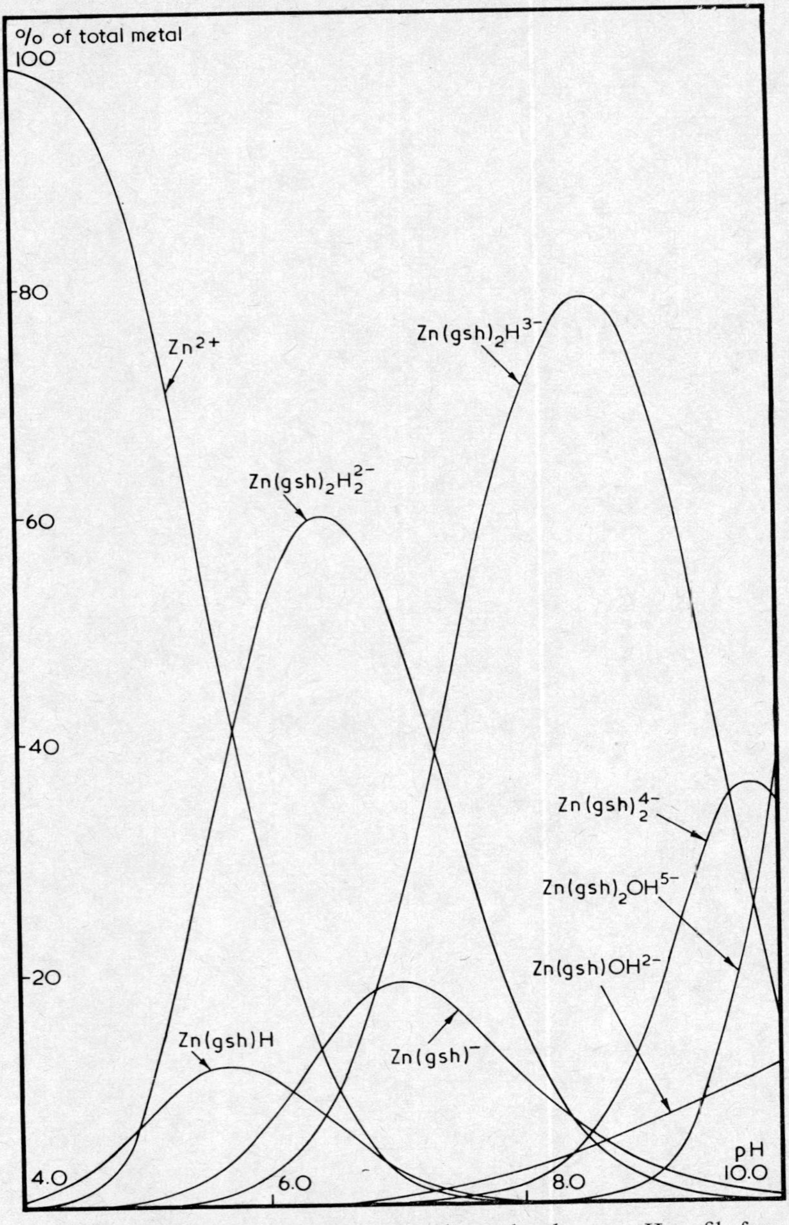

FIG. 26. An example of a percentage metal complexed *versus* pH profile for a metal complexing system. (More examples appear in Chapter 6.) The interaction of zinc(II) and glutathionate; the total concentration of zinc(II) = 10 mmol dm^{-3}, and the total concentration of glutathionate = 20 mmol dm^{-3}.

the metal ion too firmly because, once the complex has passed through the organic epithelial lining, it has to release its metal ion to the complicated mechanism that evolution has established for assimilating this metal into proteins *etc*; for example, some of the ferrous ions finish up in haem protein and some zinc ions in carbonic anhydrase or alcohol dehydrogenase. The condition (*b*) is achieved satisfactorily if the tissue protein (first stage in the assimilation mechanism) has a greater complexing ability than the ligand administered with the metal ion, *ie* $\beta_t > \beta_d$ where β refers to the overall complexing ability and subscripts t and d refer to tissue protein and drug respectively. These overall complexing ability factors are conveniently displayed as per cent metal complexed *versus* pH profiles (these are computed from formation constants and total concentration conditions)—an example is given in *Fig. 26*.

The assimilation of metals contained in food is a more complex problem. (*i*) They are often present as firmly bound complexes to food proteins (β_p). (*ii*) Labilising ligands are required that can overcome this protein binding (β_l) and release the ion into a more mobile form. (*iii*) This ligand exchange may occur in the stomach at pH = 2 or in the intestine at pH 6–7. (*iv*) The labile complex, as in the previous paragraph, ought to be lipid soluble, and (*v*) it must eventually liberate its metal to the tissue protein involved in assimilation (β_t). Thus, $\beta_t > \beta_l > \beta_p$ and we ought to remember that this complexing competition has a range of pH over which it can occur; that some of the ligands are used up by other metals and that there is a certain amount of challenge from other ligands. To quote just two examples of these complications: eggs contain iron and yet their albumin content actually decreases the amount of iron which can be absorbed from a meal; phytic acid (which occurs in cereal foods) decreases the proportion of zinc absorbed from the diet.

pH profiles of the type shown in *Fig. 26* are helping to unravel the involved interplay of complexing reactions in these processes. Finally, we must remind the reader that this sort of pattern occurs every time a metal ion crosses a membrane barrier or leaves one to be inserted into a protein—many hundreds of times in a cycle like those shown in *Figs 5* and *25*.

The biphasic response and the toxicities of essential metals

It is essential to destroy the myth that administering an element that is essential to human life 'cannot do any harm'. *Figure 2* of Chapter 1 has outlined the concept that concentrations of metals surplus to the reference values detract from the health of an organism. We can now describe some of the mechanisms through which this occurs.

Albert's description of this biphasic response principle is that: 'given too little the organism suffers severely; this is understandable from our knowledge of the large number of enzymes which cannot function without the appropriate trace metal. But, if the organism is given too much metal, a second phase of injury is seen, due to the toxic action of the excess.'

Albert found, in 1944, that 8-hydroxyquinoline (oxine) exerts its toxic effect on microorganisms by chelating heavy essential metals and forming a lethal complex. It was noted that staphylococci are killed within 60 min by 10^{-5} mol dm^{-3} oxine but 10^{-3} and 10^{-2} mol dm^{-3} oxine have progressively lower degrees of toxicity even after 24 h. Research revealed that it was the 1 : 1 oxine : Fe^{3+} complex that was lethal whereas increasing the oxine concentration progressively produced the far less toxic 2 : 1 and 3 : 1 complexes. However, (*i*) the 2 : 1 and 3 : 1 ratios can be made toxic by adding more ferric ion and (*ii*) the toxic 1 : 1 complex can be rendered harmless by other metal ions which compete with the Fe^{3+} for oxine—examples are Cd, Co, Zn, and Ni ions (Mn, Mg and Ca are not effective because they form weaker complexes to oxine than does Fe^{3+}).

By analogy, we can now see how an excess of beneficial/essential metal can be harmful. Let us assume that a 'snapshot' of just one stage in a biological mechanism shows that Fe^{2+}–(histidinate$^-$)$_2$ is involved. Clearly adding extra Fe^{2+} encourages the 1:1 complex which is of no use in the mechanism; so too Co^{2+} or Zn^{2+} metal ions will remove some of the histidinate and produce the useless 1:1 complex. Alternatively, polluting *ligands* can reduce the free Fe^{2+} concentration leading to a 1:3 complex and this useless *tris* complex can also be achieved by a lack of iron. Hence we can begin to understand how both a lack of, and an excess of, a metal can accelerate the onset of ill health.

Essential elements toxic in excess

The functioning, repair and integrity of our bodies depends upon the 18 essential elements and eight beneficial elements described in Chapters 2 and 3. However, industrialisation and 'modernisation' of established agricultural practices sometimes cause us to be exposed to greater concentrations of these elements than is desirable. Further, some malfunctioning physiological control process (for example, see our description of Wilson's disease) or a child's swallowing of a parent's iron or zinc supplementing tablets can result in a build up of an essential element. In all instances it is better to avoid or prevent these excesses rather than merely to accept treatment. However, there are still many patients who require treatment. Table 16 lists the symptoms of an excess of some of these elements and some of the drugs involved in the treatment.

Table 16. The symptoms of excessive body burdens of some essential transition metals and some of the drugs used in treating these excesses.

Metal	Symptoms	Drugs used
Manganese	Orally, not very toxic *per se* but high levels suppress iron uptake and assimilation. Manganese oxide dust in the lungs produces the symptoms of schizophrenia, Parkinson's disease, tremors and impaired muscular coordination.	$CaNa_2EDTA$, L-dihydroxyphenylalanine
Iron	Orally, it has a corrosive action in the gastrointestinal tract leading to haemorrhage, vomiting, circulation collapse and coma. Eventually iron is deposited in the tissues.	Desferrioxamine, disodium phosphate (gives insoluble iron compounds)
Cobalt	Polycythaemia and cardiomyopathy.	$CaNa_2EDTA$
Copper	It produces deposits in the liver, necrotic hepatitis or cirrhosis and eventually a haemolytic crisis.	D-Penicillamine, triethylenetetramine, 2,3-dimercaptopropanol
Zinc	Orally it causes vomiting and diarrhoea. Inhaled zinc oxide fumes gives metal fume fever (*ie* pulmonary distress, fevers and chills).	$CaNa_2EDTA$

Elemental concentration gradients *in vivo*

We distinguish here between the free and total elemental concentrations. The total concentration of an element usually remains constant whereas the concentration of its free metal ion and of its various complexes can change markedly when the metal complexing system is influenced (antagonised or stimulated) by other chemically similar metals, or when a metal ion or its complex is selectively taken through a cell membrane, or when temporal changes occur (for example, circadian concentration rhythms).

These phenomena have not been completely resolved in molecular terms but in all instances there is evidence that mixed ligand complexes play a major part. These four are now discussed.

Antagonisms and stimulations between metal ions in vivo

Optimum levels for elements in soils, plants and human tissue cannot be laid down because element concentrations (and consequently their effects) are interdependent. Although this interdependency is a common characteristic of all living species and although mixed ligand complexes are the connecting links between elements, it is not possible to generalise and say that zinc, for example, is always and solely, dependent upon phosphorus, iron and calcium. These dependencies vary with species.

Some dependent elements enhance the activity of a metal ion (*ie* physiologically they appear to increase its effective concentration) whereas others detract from its activity (*ie* they appear to mask it). This latter case is called antagonism. *Figure 27* shows Mulder's chart for trace element dependency for plant growth. When one recalls that each line implies the interaction of many complexes, the challenge of biochemistry at the molecular level is brought home. The majority of these interactions occur through localised pH effects. For example, exchanging ligands complexed to a metal ion can raise or lower the pH which, in turn can enhance or detract from the concentrations of another metal ion present (*Fig. 27a*). *Figure 28* shows an enzymic example of calcium *versus* magnesium antagonism.

Membrane transport and metal ion mobility

In the magnesium/calcium illustration just presented we can also

FIG. 27. (*a*) Mulder's chart indicating the interaction of plant nutrients showing, for example, that high concentrations of phosphates antagonise the uptake of zinc, copper and potassium whilst simultaneously stimulating the uptake of magnesium.

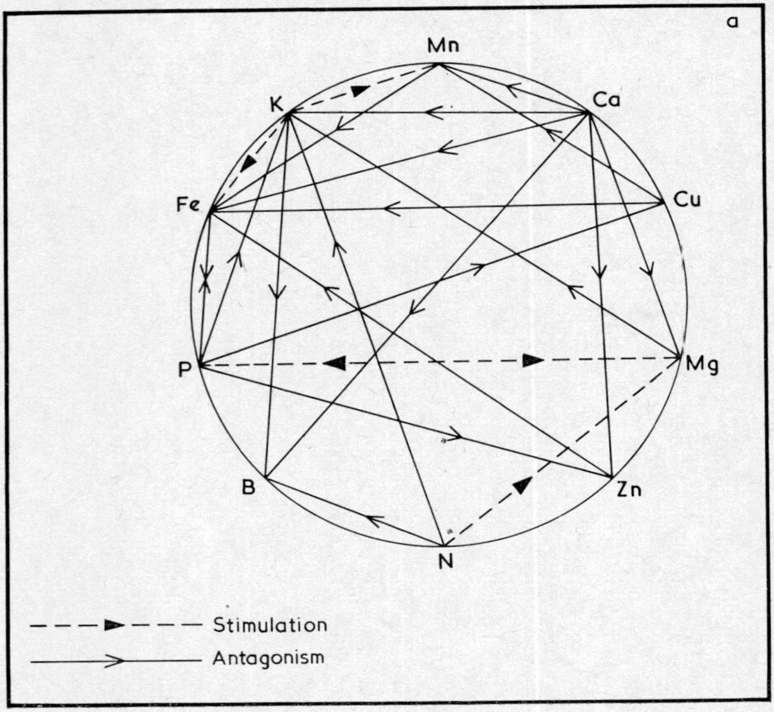

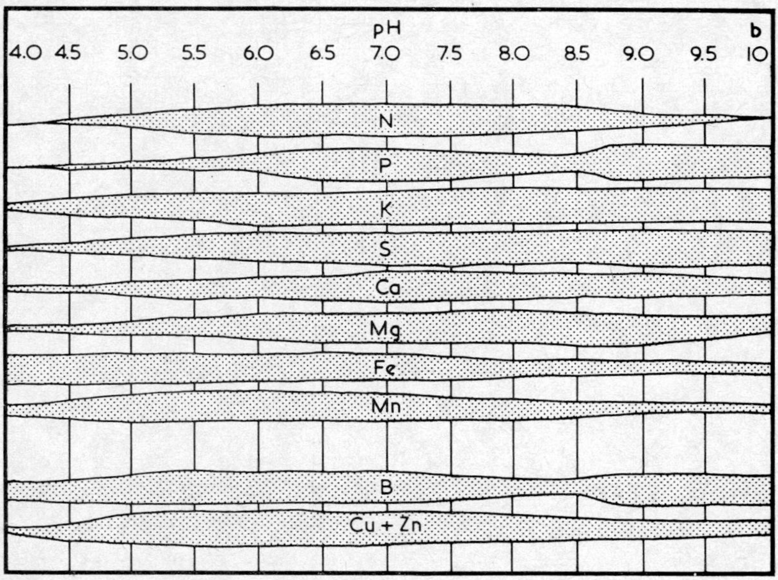

(*b*) These interactions often occur through pH relationships. This figure shows the general relationship between soil pH and the availability of elements to plants. The width of the band at the given pH indicates the bioavailability of that element.

perceive concentration gradients because of membrane transport phenomena: plasma contains about 10^{-3} mol dm^{-3} Ca^{2+} and 10^{-3} mol dm^{-3} Mg^{2+} whereas inside cells the cytoplasm concentration is far lower in Ca^{2+} (10^{-6} to 10^{-7} mol dm^{-3}) and higher in Mg^{2+} (10^{-2} mol dm^{-3}). Thus, there are both magnesium and calcium concentration gradients across cell membranes which are of opposite sign to each other. This shows the existence of mechanisms within cell membranes which maintain specific ionic concentrations on either side. Let us remember that each mechanism is a combination of many complexing steps each requiring an energy supply. Complete mechanisms are, as yet, unknown but certain intermediate steps involving ion carrier ligands have been identified.

The key concept to membrane penetration, be it for active (*ie* energised against a concentration gradient) or passive (*ie* mere diffusion) ion transport is *lipid solubility*. A membrane may be likened to a layer of organic solvent—a mixture of lipids and proteins; such solvents are not equipped to dissolve ionic salts such as those formed by the main group ions. Thus, primarily, there must be a type of carrier ligand which envelops the positive charges in a

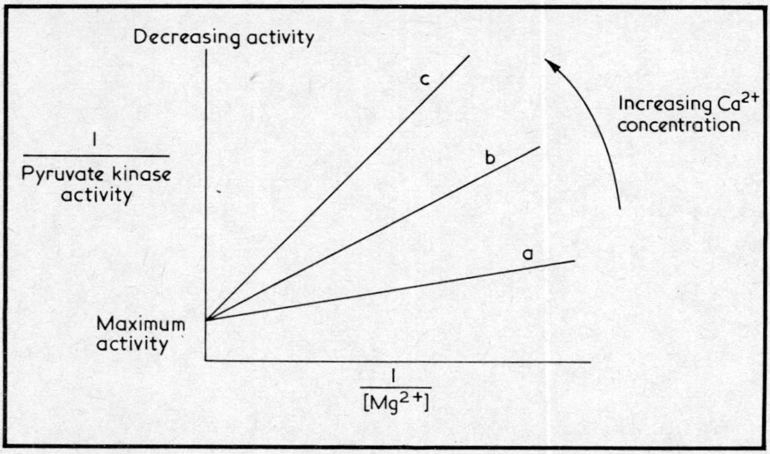

FIG. 28. An example of how calcium ions antagonise magnesium ions which are essential for pyruvate kinase activity (this is an essential enzyme in the glycolytic pathway occurring in the cytoplasm of cells). Such double reciprocal plots are useful in reducing several such experiments onto one graph. Each line refers to a fixed $[Ca^{2+}]$; $[Mg^{2+}]$ being varied. $[Ca^{2+}]_a < [Ca^{2+}]_b < [Ca^{2+}]_c$.

layer of lipophilic ligand. Further, there must be a degree of control (*a*) over which metal ion is sequestered and (*b*) over the direction in which this metal ion is propelled. We have already mentioned the Ca:Mg ratios varying inside and outside cell membranes and similar figures exist for K^+:Na^+ (10:1 inside, 1:30 outside red blood cells). These figures depend upon the sequestered metal ion being capable of fitting into a multiple enzyme process evolved for ion assimilation, ejection *etc.* We are not completely aware of the exact details of these ion pump processes although we do know that the energy necessary arises through reactions coupled to ATP hydrolysis (*see* Brown's monograph) and some ion carrier ligands have been identified. We cannot even say how these metal ion–carrier ligand complexes fit into the transport mechanism built into a membrane because we do not know the exact structure of these membranes; some workers suggest that carrier molecules stack on top of each other to form a pore lined with negative charges (from carboxylate groups on the carrier ligands) which the cation passes along, the pore being just the right diameter for a given ion and the ion being propelled against the concentration gradient by a pumping (perhaps peristaltic) movement; others favour the concept of a carrier molecule travelling through the membrane, picking up the chosen ion on one side and releasing it on the other.

Two examples of carrier ligands are shown in *Fig. 29.* Not sur-

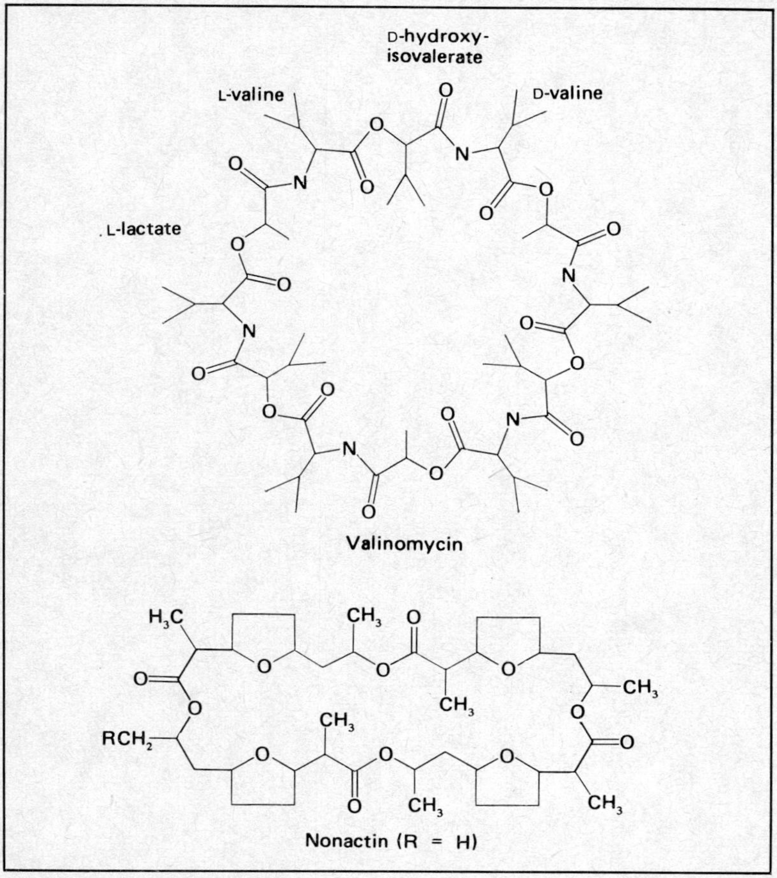

FIG. 29. Two examples of naturally occurring carrier ligands. The metal ion selectivity sequences determined *in vitro* are valinomycin ($Rb^+ > K^+ > Cs^+ > Na^+$) and nonactin ($K^+ > Rb^+ > Cs^+ > Na^+$). Note the cavity in which the cations are accommodated surrounded by ligand donor atoms and also the lipophilic exterior of these ligands.

prisingly, synthetic chemists have attempted to mimic these molecules with artificial carriers and these too have had a successful influence upon membrane transport. The selectivity for the metal ion is determined by the size of the cavity formed to accommodate the metal ion, by the number of donor groups matching up to the coordination number of the metal ion (ideally all solvent molecules ought to be replaced by ligand donor atoms), by the matching of the HSAB characteristics of the ligand to those of the metal ion, and by the presence of carboxylates which complex to form a neutral complex. If the ligand is neutral it ought to be large so that the cation's charges are spread over a large volume complex. Some of these selectivity sequences are indicated in *Fig. 29*.

Circadian rhythms

Metabolic and genetic reactions *in vivo* occur in cycles having periodicities ranging from milliseconds to months. Daily variations (circadian rhythms) occur in the concentrations of metal ions (*see Fig. 30* for an example) and of ligands. Clearly such cycles have a

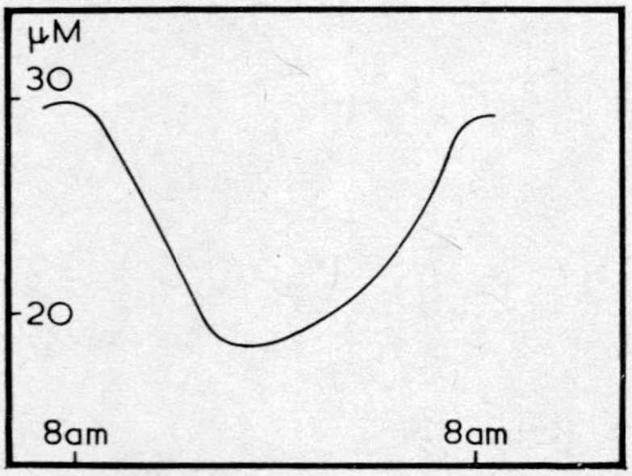

FIG. 30. The circadian variation of serum iron for a healthy human subject.

medical significance since they influence the pharmacological and toxic effects of metal complexes or ligands used as drugs. The evolutionary origin of such oscillations is unknown; perhaps they are overtones or beat phenomena of short term oscillations occurring when products of an enzyme reaction effect feedback control and modulate the degree of activity or, maybe, they are the result of our ancestral beach organisms being washed by tides twice a day or of a species being illuminated by sunlight. The biochemistry of such oscillations *in vivo* is unknown. However, the chemical principles underlying these oscillations have been investigated by Walker and Williams:

- For sustained oscillations, a continuing decrease in the free energy (ΔG) of the system is necessary; for example, humans require an input of food energy and a clock needs its spring to be kept wound up. Thus, there are some species whose concentrations are oscillating whereas others continually decrease/increase (*see Fig. 31*).
- On the one hand, the concentrations oscillate about non-equilibrium stationary states whereas, on the other hand, if the input is ceased, the equilibrium concentrations to which the system eventually dwindles is not far removed from the oscillating concentration range. This arises because nature's energy conversions are thermodynamically

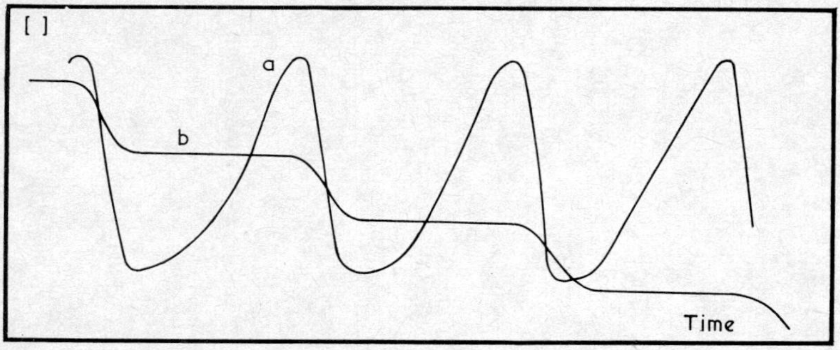

FIG. 31. Temporal concentration patterns for species involved in circadian rhythms in the body. Some species oscillate (*see a*) as the [Fe^{2+}] of *Fig. 30*, others involved in powering these oscillations vary non-monotonically and need to be replenished (*see b*), for example, food.

efficient and so occur close to reversible equilibrium concentrations (thus, we are justified in discussing equilibrium concentration solutions in the next chapter and in the section about computer models in Chapter 4).

- The amplitudes and periods of circadian rhythms can vary with the age, health and degree of pollution of the organism; for example, both ethanol and EDTA reduce the period and amplitude of reactions in model studies whereas Mn^{2+} ions cause an increase in both. Paradoxically, 'Aspirin' or Zn^{2+} ions cause an increase in period but no change in amplitude.
- As may be expected from the fact that enzymes are involved in these mechanisms of oscillation *in vivo*, at least one part of this mechanism must involve a feedback loop.

A whole new area of therapy based on optimum timing of drug administration (chronotherapy), making due allowance for rates of absorption, metabolism, excretion, the amount of food in the stomach and other factors, has recently sprung up because circadian rhythm influences are not trivial; for example, in principle, 'desferrioxamine B' treatment to remove excess Fe^{2+} in serum can be 65 per cent more effective when administered to coincide with the maximum concentration in serum (*see Fig. 30*) compared to the minimum concentration: this percentage increase can often represent the difference between the success and failure of a therapy.

Mixed-ligand complexes

Chapter 4 has described the intriguing coordination chemistry of mixed-ligand complexes. However, it remains for this chapter to underline the complete dependence of living systems upon such

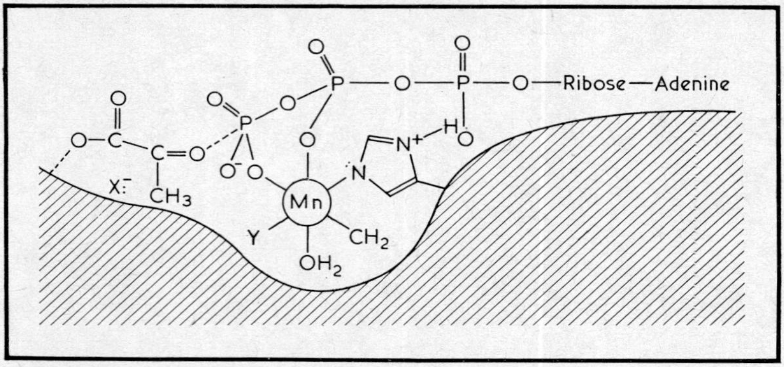

FIG. 32. An example of a mixed ligand complex—the active site of pyruvate kinase. Note how many different ligands are complexed to the Mn^{2+}.

complexes. A listing of the roles of mixed ligand complexes should make this principle self-evident; they are necessary for the proper functioning of the enzymes (specificities of both metal activated and metalloenzymes involve entatic states which are a direct result of mixed ligand complexing (*Fig. 32*)), in the invasion of cells by viruses (*Fig. 33*), in the storage and transport of metal ions (*Fig. 34*), in the membrane transport just discussed, in the process of nitrogen fixation and the transport of oxygen in the blood, and in the complex competition between roots, soil particles and organic ligands (for example humic acids) through which metal ions enter plants—schematically shown in *Fig. 35*.

FIG. 33. A diagrammatic representation of a virus invading a cell. Note that both the connection between virus and cell, and the rupturing of the viral sheath, to permit the viral DNA to enter the host cell, involve mixed ligand complexes of Zn^{2+}.

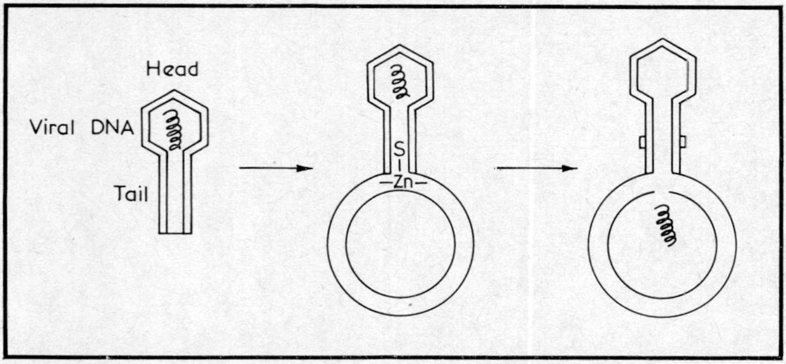

$$Cu^{2+} + \text{amino acid anion} \rightleftharpoons Cu^{2+}\text{—amino acid complex}$$
$$Cu^{2+}\text{—amino acid complex} + \text{albumin} \rightleftharpoons \text{albumin—}Cu^{2+}\text{—amino acid complex}$$
$$\text{albumin—}Cu^{2+}\text{—amino acid complex} \rightleftharpoons \text{albumin—}Cu^{2+} + \text{amino acid anion}$$

FIG. 34. Mixed ligand complexes are used to conduct a free cupric ion *via* a Cu^{2+}-amino acid complex into its binding site in a blood protein called albumin.

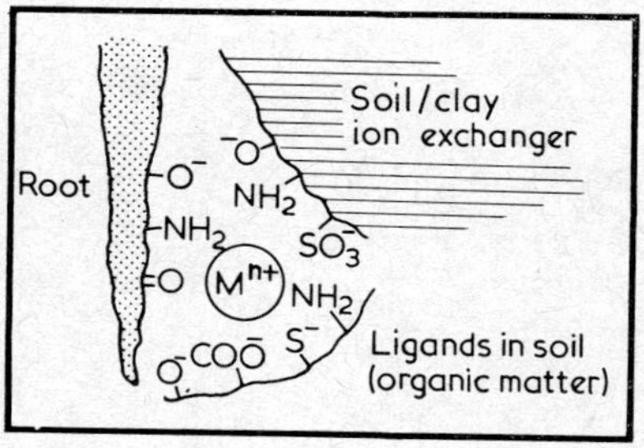

FIG. 35. Diagrammatic representation of the competition between roots, soil and ligands for a plant nutrient metal ion. The root either wins its metal ion directly from soil particles, or decaying vegetation produces organic ligands which remove the metal ion from the soil and then release it to the more powerful ligand donor groups of the root.

It should be clear that nature is considerably more advanced in mixed complex terms than the symmetrical octahedral $[Ni(H_2O)_6]^{2+}$ examples discussed in undergraduate courses; instead of symmetry we find asymmetry and bond strain; instead of six equivalent ligands we find the involvement of as many as six different donor groups and sometimes two or more metal ions. There is a very good evolutionary justification for such complexity: it means that complex *homo sapiens* and his environment have been able to produce millions of different complexes from relatively few building blocks (ligands). Since viable synthetic steps had to be evolved for each of these ligands, a considerable time saving has been witnessed on the evolutionary scale by keeping their number as few as possible.

The aetiology of pathological states

The aetiology of selenium dependent states

In Chapter 3 we mentioned that selenium is beneficial to health because it is related to the protection against the oxidative effects of hydrogen peroxide. This is about the sum total of our *chemical* knowledge of the bio-inorganic aspects of selenium. On the other hand, there is much aetiological evidence implicating selenium with disease and this sorely needs molecular level investigation. However, the role of selenium in health and disease is multifarious. At a level greater than 5 μg per gram of diet in animals it is very toxic, producing liver necrosis, exudative diathesis, muscular dystrophy, 'blind staggers' and 'alkali disease' as found in livestock in the mid-west United States. This poisoning occurs through the accumulation by plants of the *Astragalus* species of concentrations of selenium which exceed the tolerance level of grazing cattle. Basically, the symptoms are hair and hoof loss presumably caused by selenium replacing sulphur in sulphur containing amino acids in keratin (*see* later). On the other hand, a dietary content of up to 3 μg(g of diet)$^{-1}$ improves growth and combats white muscle disease in sheep. Selenium deficiency causes the enzyme glutathione peroxidase to exhibit reduced activity (selenium might conceivably be an active part of this enzyme). Non-protein selenium is more *acutely toxic* than protein selenium whereas the latter is more *chronically toxic* than the former. Paradoxically, the presence of selenium in tuna fish *protects* the consumer against mercury poisoning.

Selenium has a cancer protecting effect:

(*a*). Blood plasma samples taken from patients with cancer have selenium counts which are lower than the reference value. This study has been augmented by analysing the selenium content of blood from 20 US blood banks and finding that cancer mortality rates area for area increased as blood bank selenium contents decreased.

(*b*). Blood selenium originates in diet and a similar cancer protecting effect of an optimum value of selenium in food grown locally (*eg* potatoes and grain) has been observed from analyses in 50 of the states in the US, there being a higher incidence of cancer in states too low or too high in selenium (a manifestation of *Fig. 2*). Similarly, areas of Canada which grow selenium indicator plants have lower cancer mortality rates.

There are some fascinating observations concerning the selenium status of animals in Turkey—serum selenium levels exhibit seasonal surges that differ in timing between sheep and goats, *although both graze the same common pasture*. Since selenium deficiency leads to infertility and since the breeding periods of sheep and goats differ, Hemsted has suggested that the selenium status of the animal might be dependent upon seasonal hormone changes as coat shedding is also seasonal. It is postulated that the cancer sensitivity of humans to selenium might be because we have lost our seasonal reproductive traits and annual coat shedding mechanism but that latent forms of this mechanism can still be selenium stimulated into life leading to hormone based cancers and skin cancers respectively.

Clearly both the cancer protecting effects of selenium and the aetiology of selenium related pathological states, need further investigation as also do the chemical mechanisms behind these fascinating observations.

Selenium is by no means unique in having a weight of circumstantial evidence connecting it with cancers. Other elements currently under suspicion are cadmium, copper, magnesium, molybdenum and zinc.

Molybdenum dependent fluorosis

The construction of a large dam in Southern India has raised the water table in the surrounding area; this has made the top-soil more alkaline and upset the trace element balance for the local peoples, leading to *genu valgum* (the crippling bone disease leading to knock knees and concomitant emotional disturbances and social stress) in young people.

Apparently, the increased soil alkalinity provides suitable chemical conditions for sorghum plants to take up molybdenum (these plants are the staple diet of the indigenous races) and fluoride. These, in turn, lead to a high copper excretion rate (threefold). In experimental animals copper deficiency produces similar osteoporosis and so, clearly, it is the interaction of the two elements copper and molybdenum in human nutrition that causes *genu valgum*. Clearly, epidemiologists and nutritionists must strive to understand more of the importance of trace elements in the diet of borderline populations and chemists must aim to unravel some of the complicated mechanisms involved.

Cancer distribution

Many diseases are dependent upon the trace-element composition of the soil in which food-crops are grown—the US has a goitre belt related to the low concentration of iodine, and in areas of high selenium content there are many cases of gastric, intestinal and hepatic dysfunctions. Similarly, multiple sclerosis and such neurological diseases occur in areas of elevated soil lead content. As far as malignant disease is concerned, the Netherlands has a high silicon content which is reflected in a high incidence of all cancers. However, when sufficient calcium is present this increased danger is averted. Indeed, in the UK chalk and limestone areas have low cancer rates whereas high zinc and chromium soil contents have been associated with gastro-intestinal cancer. In South Africa, molybdenum, copper and iron deficiencies have led to oesophageal cancer.

Besides soil and plant constituents being sources of excesses, industrial food processing can also elevate concentrations; for example, abrasions on equipment can lead to As, Pb, Cu, Zn, Ni, Hg and Sb being found in food and, of course, metals such as nickel are used as catalysts in the hydrogenation of oils to fats.

The case of magnesium soil deficiencies bears closer scrutiny. In France, the low magnesium soil content is associated with a greater general incidence of cancers and this phenomenon has been further investigated by Aleksandrowicz in Poland. Some regions of Poland had a very high incidence of leukaemias. Fortunately for the purposes of his investigation, many Polish families eat food grown in their own gardens and this has enabled the Aleksandrowicz team to perform soil analyses, fungal analyses and food-from-the-plate analyses to establish the identity of the cancer causing or promoting agent.

They found that occurrence of tumours and leukaemias in humans in Poland correlated with the contamination of dwellings, occupied by patients for many years, by some *Fungi imperfecti* known to have oncogenic properties. Tumours of the liver and haematological disorders could be induced in carp fish by means of these fungi or by their metabolites that had been isolated from the dwellings of cancer patients, propagated under laboratory conditions and then administered orally to the fish. Further, the sera of patients with proliferative haemocytopathies were found to contain antibodies against *Aspergillus flavus* and other fungi which contaminated their dwellings. Finally, the homes of these cancer patients occurred in clusters and were found to be contaminated with *Aspergillus flavus* and other oncogenic fungi. Such groups of houses could be associated with an excess of silicon and potassium, and a deficiency of iron, copper, and magnesium in drinking water or soil. This all suggested that, under certain trace element environmental conditions, fungi known to have oncogenic properties could promote leukaemic processes and other disorders of the haemopoietic system in man.

There was also a further related finding—tumour incidence was far higher (up to 10 times the incidence) in areas where potassium, nitrogen and phosphorus fertilisers had been distributed by previous occupants.

What can chemists contribute to this detective story? Firstly the organic or biochemist can identify the metabolic products of *Fungi imperfecti* that are the actual carcinogens. This was not such a difficult task because there had been ample precedents; mycotoxicoses (diseases resulting from ingesting toxins produced by fungi or moulds growing on foods) were first identified about 80 years ago and approximately every decade since then there has been a reawakening of interest; for example, in the 1950s in Japan, mycotoxins were associated with the aetiology of liver cancer and then in 1960 there were the tragic losses in Britain from Turkey X disease caused by a product of *Aspergillus flavus* from Brazilian groundnut which was used as a foodstuff. This product was identified as aflatoxin (*Fig. 36*). Subsequently, it was discovered that it was not just a carcinogen but *the* most potent carcinogen known to man. Then in the 1970s there were these Polish discoveries.

Humidity and temperature clearly play a part in encouraging fungal growth but there is very definitely also a top soil metal ion

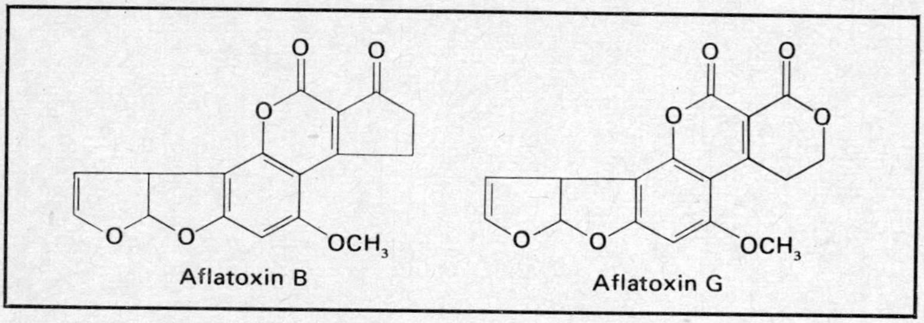

FIG. 36. The chemical formulae of some aflatoxins.

composition effect. It is here that the bio-inorganic chemist can present his postulates. Aleksandrowicz's studies have suggested that the following ions are implicated—Mg^{2+}, Mn^{2+}, Si^{4+}, K^+, $Fe^{2+/3+}$, Se^{2-}, Zn^{2+}, Cu^{2+} and Mo^{2+}.

Here we refer the reader back to our discussion of *Fig. 2* and point out that such a concept of optimum concentrations applies to humans in health, to tumours growing in humans and to *Fungi imperfecti*, the optimum magnesium concentrations for healthy growth differing between these three species. Three possible schemes resulting in carcinogenesis are shown in *Fig. 37*. We might note that each of these arrows could have side branches or amendments to allow for the 3rd axis in *Fig. 38*, *ie* the concept that the Mg:Ca and Mg:K ratio is more important than the absolute concentration. This could explain the potassium fertiliser influence. Yet further complications to the figure would arise, (*i*) from the possibility that it is not aflatoxin itself but rather a human metabolic product that is the carcinogen, (*ii*) from the question of whether this carcinogen is an initiator or a promoter of cancer and (*iii*) from the question of whether the aflatoxin binds directly to host cell DNA or through a metal ion as an intermediary.

FIG. 37. Possible schemes of carcinogenesis in humans involving metal ions, fungi and aflatoxins.

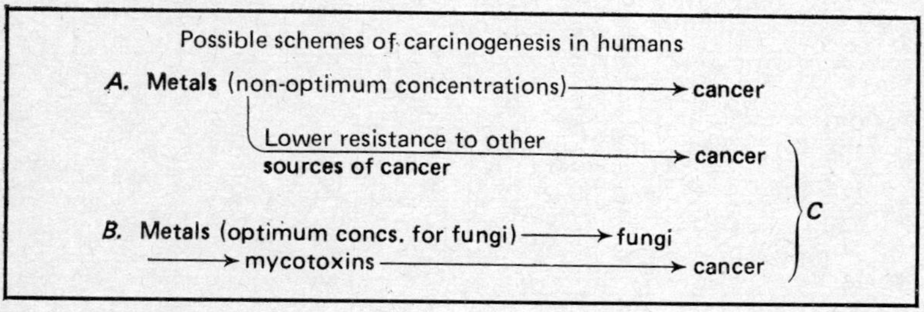

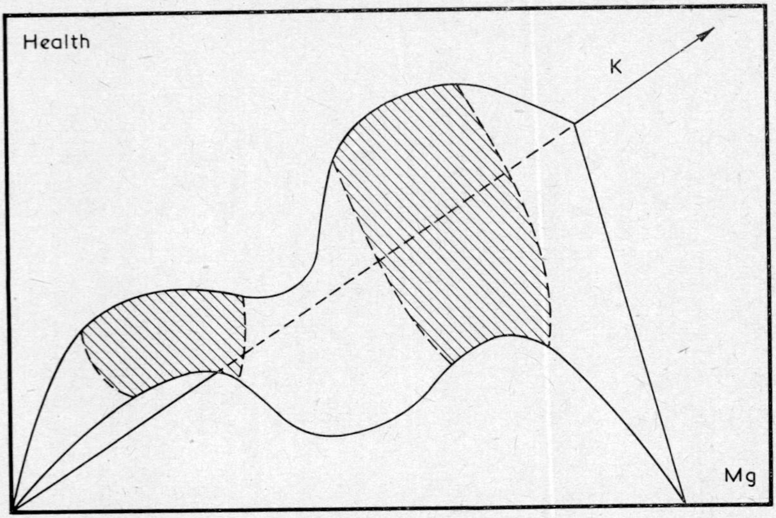

FIG. 38. This is a more sophisticated version of *Fig. 2* and is a three dimensional representation of how the ratio of K: Mg concentrations determines the health of a fungus that produces aflatoxins, of a human or of his immune defence system. Thus soil fertilisers containing K can influence the biological effect of different Mg concentrations.

At all of these stages we must realise that whenever organisms (fungi, plants, humans, *etc*) are placed under physical stress, one or more of the elements may become growth limiting or encouraging. Thus many more elemental analyses of soil, food, water, tumours, *etc* and more molecular studies are necessary before the precise principles of magnesium deficient carcinogenesis can be established.

However, the following prophylactic/therapeutical measures have already been implemented: (*i*) Dolomite, $CaMg(CO_3)_2$, is spread on the land to prevent the growth of *Fungi imperfecti*: (*ii*) Magnesium is necessary in man to enable the immune response system to function so that invading organisms can be repelled. The soil and water in north west Poland (an area previously inhabited by German farmers who used large amounts of K, N, P fertilisers) have magnesium contents which approach zero levels and so diets are being supplemented by replacing vacuum pan salt (>99 per cent NaCl, <0.03 per cent $MgCl_2$) with rock salt (for example, 97 per cent NaCl, 1 per cent $MgCl_2$). There are many different types of the latter in Poland and there is an ongoing project to choose the salt having an ionogram most closely related to the requirements of the human diet. Since our primitive cellular ancestors evolved in the oceans and on the beaches and so bodily fluids such as blood, sweat, and tears are salty, it seems logical to supplement our diets with the salts from ancient oceans. (*iii*) Magnesium levulinate is administered intravenously to leukaemia patients until their erythrocyte magnesium levels are normalised or until the Ca : Mg and K : Mg ratios in body fluids are within 95 per cent of normal.

Magnesium and potassium are, of course, essential elements and we must remind ourselves that polluting elements can also challenge normal health (for example, chromium and gastro-intestinal cancer). However, probably because of the scientist's conscience at having released these elements into the biosphere, there has been a great deal of effort put into understanding the *modus operandi* of the pollution process and, indeed, in chemically designing ligands to remove these pollutants. There are many important principles involved in this subject which will be discussed in the following chapter.

6. The Principles of Bio-inorganic Medicine

'If bio-inorganic chemists are so clever at producing drugs, why am I still sick?'

This far reaching question is best answered in two parts.

(*a*). The process of growing older involves our bodies becoming more polluted with exogenous and endogenous chemicals—the gifts of our polluted environment and the products of our metabolism. Elderly people have chemistries which are slower to react to the challenge. We must also remember that some symptoms (*eg* pyrexia, urticaria, catarrh, vomiting and diarrhoea) are actually a figment of our defence mechanisms. These manifestations of ageing are not likely to yield to new cures in the more immediate future although there may be new treatments of these conditions which reduce the pain and torment of these situations which, in the main, arise from our in-built (*ie* genetic) obsolescence.

(*b*). There are many other conditions which can be treated, and often cured, by operating the principles described with reference to *Fig. 2*. Clearly, bio-inorganic scientists can contribute in at least four areas—(*i*) ligands to remove polluting/contaminating metal ions; (*ii*) ligands to remove beneficial/essential metal ions present in excess; (*iii*) metal complexes to supplement deficient essential/beneficial elements; and (*iv*) metal complexes to stimulate the host's defence mechanisms. The areas of greatest success are those in which one organ or function of the body is out of step—as distinct to an overall degeneration of standards as in (*a*). In these former patients, metals, ligands and metal complexes used as drugs are enjoying a great deal of success. This chapter describes a range of these bio-inorganic agents and the principles involved in their selection, administration, dose regimens, and mechanisms of activity. Finally, we suggest some future trends.

But first, it is instructive to see the history of drug discovery outlined.

The evolution of modern therapeuticals

The history of drug use can best be summarised in tabular form—Table 17.

Healing has been practised for many centuries, the earliest report of metallotherapy being from 1500 BC when an aqueous–ethanolic suspension of rust was administered as a cure for impotence. In the Greek and Roman period scientific method was first introduced into

Table 17. The history of drug use.

Dates	Country	Drugs	Disease treated	Notes
1600 BC	Egypt	Plants, minerals	All ailments	In 1500 BC iron in the form of rust was used to treat impotence (Melampus).
430 BC	Greece	Herbal remedies	All ailments	Drugs were based upon the case history of the patient (Hippocrates)
2nd century	Asia Minor	Mutiple herbal remedies	All ailments	Drugs were based upon the four 'elements'—fire, air, water and earth (Galen)
16th century	Switzerland	As, Sb, Hg, Cd, Fe	Syphilis, anaemia	Condemnation of herbal mixtures; optimised drug doses (Paracelsus)
18th century	South America	Cinchona alkaloids	Malaria	Digitalis isolated (Withering)
19th century	France, UK	Curare *etc.*	Analgesics	Drugs first separated into ethical medicines (formula was published) and non-ethical or proprietary medicines (ingredients were a closely guarded secret) (Barnard *et al*)
1899	Germany	Aspirin	Analgesic and antipyretic	C, H, O drugs as offshoots of the German dye industry (Bayer Co.)
1903–1910	Germany	Veronal, Salvarsan	Insomnia, syphilis	(Ehrlich)
1920	Canada	Insulin	Diabetes	(Banting and Best)
1928	UK	Penicillin	Microbial infections	Unrecognised until 1940 (Fleming)
1935	Germany	Prontosil	Streptococci	S, N drugs, the first sulphonamide (Domagk)
1943	UK	M and B 693	Pneumonia	Saved a great many lives
1944–	UK	Many	Broad spectrum of diseases	Rapid expansion in the drug industry (In 1970, 70 000 employed in the UK drug industry)
1950s	Australia	Anti-viral metal complexes	Viral infections	(Dwyer)
1964–	US	Platinum complexes	Cancer	(Rosenberg)

medicine by Hippocrates who kept detailed case histories and used this information to decide the quantity of drug to be administered. Paracelsus (1493–1541) made the important observation that all substances in sufficient quantities are toxic, even water in excess is fatal, and so introduced regulated doses into therapy.

As little as 100 years ago, most drugs were crude preparations, obtained from plant, animal or mineral sources, few of which were specific or even effective in relief of symptoms. However, at the end of the nineteenth century there were great advances in organic chemistry and also in the fields of bacteriology, pharmacology and immunology. Around this time important steps forward were Louis Pasteur's theory of infection and Lister's achievement of antiseptic surgery.

Even in 1930 only six specific drugs were known; these being quinine for malaria, ipecacuanha for dysentery, digitalis for heart conditions, mercury for syphilis, 'Salvarsan' for syphilis and vitamins for specific deficiencies. Then, however, the drug revolution really got underway, when the sulphonamides were discovered as a spin-off from the German dyestuffs industry, until today more than 30 000 drugs are available, some 8000 of which are commonly prescribed.

Apart from the reasonably widespread use of Na^+, Ca^{2+} and Mg^{2+} salts for intestinal treatment, Fe^{2+} salts for the treatment of anaemias and some use of mercury and arsenic for syphilis and trypanosomes, metallotherapy has not developed at a rate comparable with our understanding of the roles of metals *in vivo*. The connection between toxic action and metal coordination was first proposed by Voegtlin in 1923 when he suggested that the toxic action of arsenic on living cells was due to its combination with essential thiol compounds. There are an enormous number of drugs which can bind metal ions and so, presumably, such binding occurs *in vivo*.

Metallotherapy advanced again in the 1950s when the antiviral properties of metal complexes were discovered, and, as more cancers are found to be virus dependent, metallotherapy has entered into the anti-cancer battle.

Today there are four basic ways in which the search for a new drug can be carried out: (*i*) by isolating, analysing and imitating natural products; (*ii*) by schematically varying the molecular structure of existing drugs, *ie* 'playing molecular roulette'; (*iii*) by the random screening of all chemicals; (*iv*) by a theoretical approach upon a physiological basis. The last of these methods would, obviously, be the most rewarding, but it can seldom be used because of our lack of knowledge of biological processes at the molecular level as discussed in Chapter 4.

The actual term 'chemotherapy' was introduced by Ehrlich at the turn of the century and he defined it as 'the use of drugs to injure an invading organism without injury to the host'. The underlying principle of therapy using chemicals is the establishment of differential effects whereby the patient gains some advantage over disease through the administration of the drug. Often, this advantage is one of bolstering up the person's natural defensive forces and tipping the balance in his favour. However, we must recognise that the majority of drugs have some disadvantages—nephrotoxicity, addiction, tiredness, *etc*. Medical practitioners are skilled in the art of balancing the advantages against the disadvantages. Their decisions will eventually be made easier by quantitative computerised medication (*see* later) as distinct to all adult patients, regardless of age, weight or sex, being prescribed 'a 5-ml dose three times a day after meals'.

Ehrlich expressed the state of this balance as a chemotherapeutical index (CI) which was described as minimal curative dose ÷ maximum tolerated dose. This is synonymous with the minimum dose to achieve the advantageous reactions ÷ maximum dose for which the disadvantages can be tolerated by the body. More recently, this chemotherapeutic index has been modernised by inversion and definition as LD_{50}/CD_{50}, *ie* lethal dose in 50 per cent of the test animals ÷ dose curing 50 per cent of animals. Such a concept underlines the principle that it is pointless to market a new agent with half (say) the toxicity to the host unless it is more than half as active as the established drug. Some examples of CI are given in Table 18. Trevan has pointed out that the intensity of cellular

FIG. 39. Cellular responses to a drug dose show a Gaussian distribution.

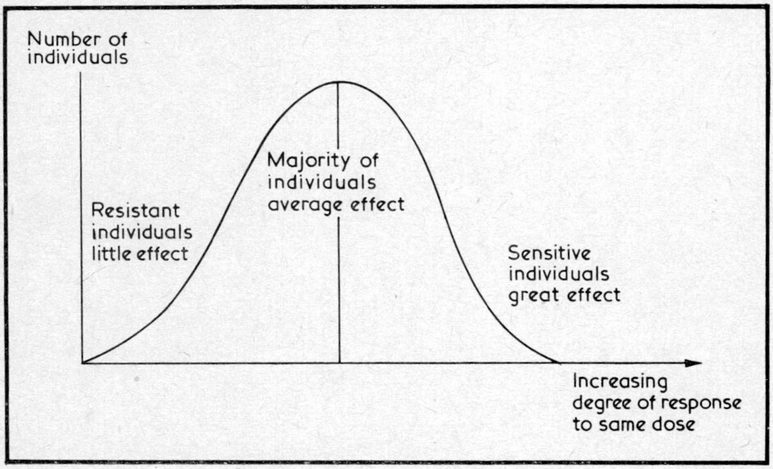

responses to a specified dose of drug will usually show a Gaussian distribution (*Fig. 39*).

Table 18. Some examples of chemotherapeutic indices.

Drug	CI
Local anaesthetic	
'Cocaine'	53
'Procaine'	25
Antihistaminic	
'Pyrilamine'	68 000
'Hetramine'	500
Anticancer	
Malonatodiammineplatinum(II)	12.2
Cis-dichlorodiammineplatinum(II)	8.1
1,2-Diaminocyclohexanedichloroplatinum(II)	6.9
Cis-dicyclohexylaminedichloroplatinum(II)	270

Means of administering chemicals to humans

The routes of administration of a drug may be separated into two general classes—enteral and parenteral. For enteral administration the chemical is placed directly into the gastrointestinal tract by permitting a tablet to dissolve when placed under the tongue (sublingual), by swallowing a tablet or a medicine (oral) or by rectal administration of tablets or capsules. In parenteral administration the gastrointestinal tract is not used. Such routes include subcutaneous, intramuscular and intravascular injections, topical applications such as skin ointments, pessaries or aerosol inhalations.

There are many factors taken into consideration when deciding a route of administration. These include convenience to patient (*eg* self-injection is generally not feasible), the shelf-life of the chemical (*ie* solid *versus* solution), metabolism of the agent *en route* to its site of activity, the rate of onset of and duration of the drug response (on the one hand intravascular administration is immediate whereas sustained release capsules, implants, *etc* are effective over a far longer period), body tolerance and side effects (irritation, pain, *etc*), the possibilities of abuse (overdoses, addiction), and the range of cellular membranes that the drug and its reaction products must traverse during its residence and activity *in vivo* (these considerations are dependent upon the pH of the aqueous solutions bathing each side of these membranes and on synergistic reactions between drugs).

Clearly, we cannot describe all of these factors in detail for all metals, ligands and complexes administered as drugs (indeed in many instances they have not been fully characterised; thankfully, the safety aspects of any unknown grey areas are always checked out by screening experiments before a drug is marketed). However, we will describe examples of some of the more salient principles.

Absorption of ligands from the gastrointestinal tract

If at all feasible, the oral administration of a drug is obviously the most convenient method. In general, enteral absorption occurs throughout the whole length of the gastrointestinal tract but in three localities appreciably more intense absorption may occur depending upon the dissociation constants of the ligand. These three areas are the mouth (pH of bathing solution $\simeq$ 7.4), the stomach (pH $\simeq$ 1.6) and the small intestine (pH (duodenum) $\simeq$ 6–6.5; pH (jejunum) $\simeq$ 6.5–7).

Specific carrier systems exist for the transport of some species across the intestinal wall, *eg* glucose or amino acids. For other species, however, the properties required for good absorption are the presence of a high proportion of a non-ionised form with a high lipid:water partition coefficient and a small atomic or molecular radius. It is generally assumed that the ionised form of a weak acid or base cannot cross the mucosal lining of the gastrointestinal tract but that the non-ionised form equilibrates freely.

If there is a pH difference across the gastric mucosa then the fraction of a weak acid which is ionised may be considerably greater on one side than on the other. Thus, at equilibrium, when the concentration of the non-ionised form is the same on both sides, there will be more drug in total on the side where the degree of ionisation is greater.

Consider, for example, how the weak acid, salicylic acid, distributes across the gastric mucosa between gastric fluid (pH $\simeq$ 1.6) and plasma (pH $\simeq$ 7.4).

A weak acid dissociates according to the equation

$$\mathrm{AH} \rightleftharpoons \mathrm{A}^- + \mathrm{H}^+$$

and the extent of dissociation is described by the equilibrium constant $K_a = [\mathrm{AH}]/[\mathrm{A}^-][\mathrm{H}^+]$. Thus we see that the extent of dissociation will depend on $[\mathrm{H}^+]$, *ie* pH.

For salicylic acid, $\log K_a = 2.8$ and *Fig. 40* shows the percentage of each form which is present as the pH is changed. At stomach pH 1.6 we have 6.5 per cent A^- and 93.5 per cent AH and at plasma pH 7.4 we have 99.99 per cent A^- and 0.01 per cent AH.

Thus, if no transfer occurred across the stomach wall, we would have the situation as shown in *Fig. 41a*. However, when transfer of the non-ionised form occurs its concentration becomes the same on both sides and so we have the situation as in *Fig. 41b* and so we see that most of the ligand is transferred from the stomach to the plasma.

Thus a weak acid, which is largely in the non-ionised form at

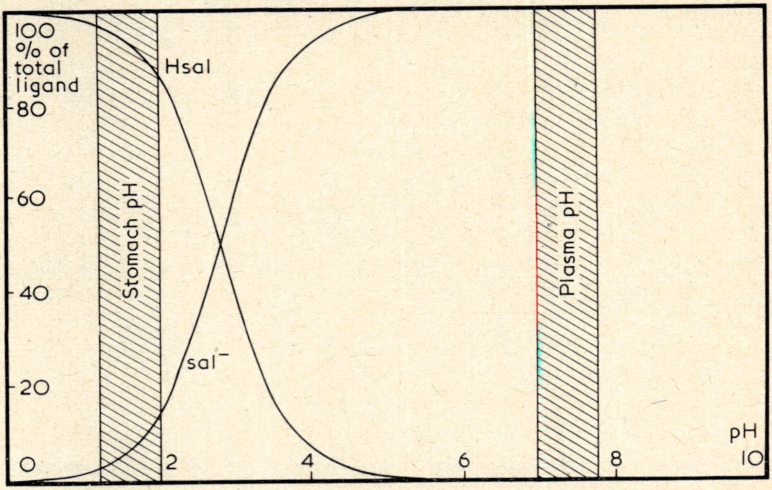

FIG. 40. The pH dependence of the form of salicylic acid.

pH 1.6, is readily absorbed from the stomach. On the other hand, at this pH a weak base is mainly in the charged form BH^+, and is not readily absorbed.

However, we must remember that as soon as a ligand is absorbed across the stomach wall it is carried away by the blood and so simple reversible equilibrium across the wall does not occur until the ligand is distributed throughout the whole bloodstream.

Absorption of metal complexes from the small intestine

The same general principles apply as above, *ie* only non-ionised

FIG. 41. (*a*) The distribution of salicylic acid assuming no transport across the stomach wall.
(*b*) The equilibrium distribution of salicylic acid when transport across the stomach wall occurs.

Plasma pH = 7.4		Stomach pH = 1.6	
$H^+ + A^- \rightleftharpoons HA$ 99.99 0.01		$HA \rightleftharpoons H^+ + A^-$ 93.5 6.5	a
Total ligand ~100		~100	
$H^+ + A^- \rightleftharpoons HA$ 199.98 0.02	$\rightleftharpoons$	$HA \rightleftharpoons H^+ + A^-$ 0.02 0.0014	b
Total ligand ~200		~0.02	

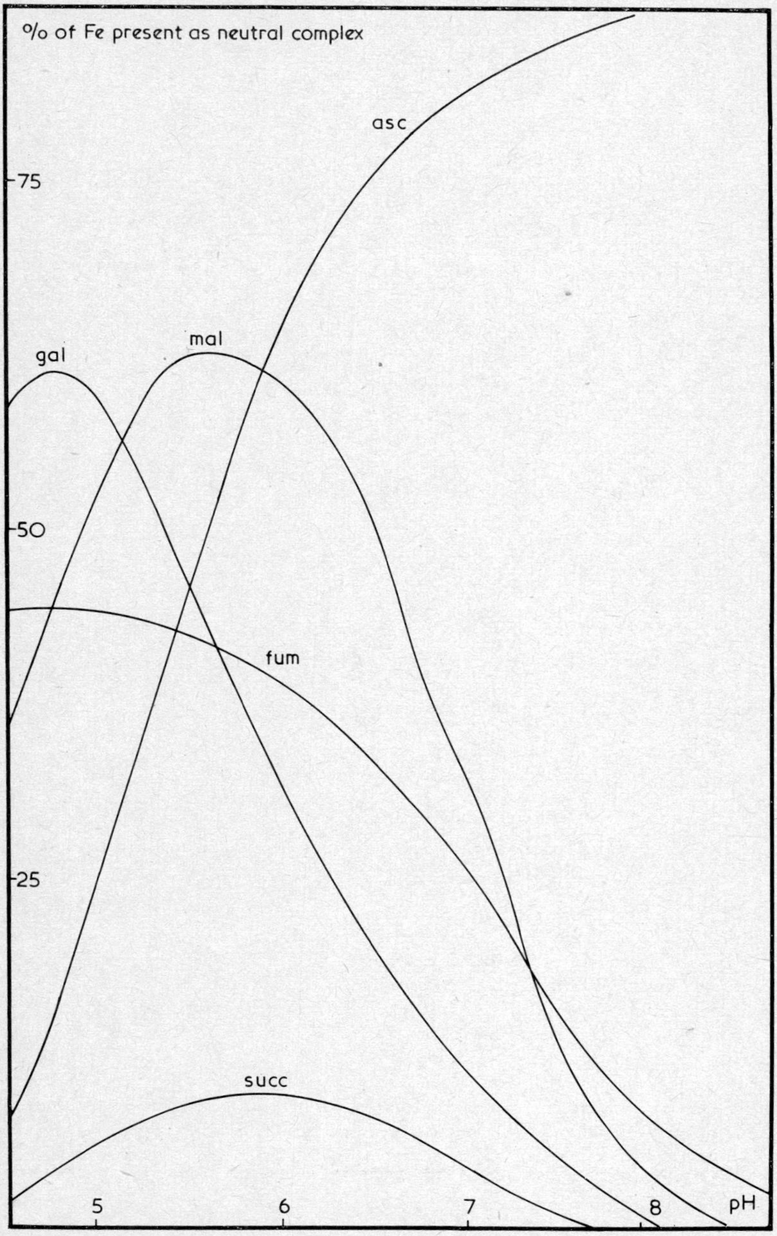

FIG. 42. The pH dependence of the concentration of neutral complex formed on the interaction of Fe(II) with ascorbate, malate, fumarate, galacturonate or succinate.

species can cross the intestinal wall, as long as they have molecular radii of less than about 30 Å which corresponds to a molecular weight of about 60 000.

As discussed in Chapter 3, iron is an extremely important metal in human metabolism and imbalances in its metabolism are fairly common resulting in iron deficiency anaemias. This condition can be treated by the oral administration of salts of Fe(II), these being partly absorbed in the duodenum and jejunum. A wide variety of ligands have been used in place of sulphate to try to increase the proportion of Fe(II) that is absorbed in the small intestine, or to decrease the amount of gastrointestinal irritation caused by the administration of ferrous sulphate.

One criterion for choosing the 'best' ligand for promoting Fe(II) absorption is that it should give a high proportion of a non-ionised complex at small intestinal pHs. If we know the equilibrium constants for the formation of all the complex species then we can determine the amount of non-ionised species present at any pH. *Figure 42* shows the concentration of neutral species formed on the interaction of Fe(II) and the ligands ascorbate, malate (hydroxybutanedioate), fumarate (*trans*-butenedioate), galacturonate and succinate (butanedioate), some of which are currently in use in iron supplementing drugs. In *Fig. 43* the percentage of neutral species present between pH 6 and 7 is shown for each of the above five ligands and so we see that we would expect their Fe(II) absorption promotion properties to be in the order ascorbate > malate > fumarate > galacturonate > succinate.

Absorption from implanted metals

Dissolution of metal from an implant may be either desirable or undesirable and so the metal must be chosen accordingly.

An example of the former case is the use of a copper 7 or a copper T as an intra-uterine contraceptive device. The use of trace elements to regulate fertility is not a new approach, copper salts having been known to be toxic to sperms since 1853.

A typical copper T IUD has a copper surface area of 135 mm^2 and releases about 29 μg of copper per day into the uterine fluid. This dissolution of the metal is much less than the 2–4 mg per day which must be taken up from the diet to maintain a normal copper balance, and the body's homeostatic mechanisms ensure that the plasma copper levels are not allowed to rise above their normal values.

Little information is yet available about the mode of action of these IUDs but it is believed that their contraceptive effect is due not to an inhibition of fertilisation but to an interference with the implantation of the fertilised ovum.

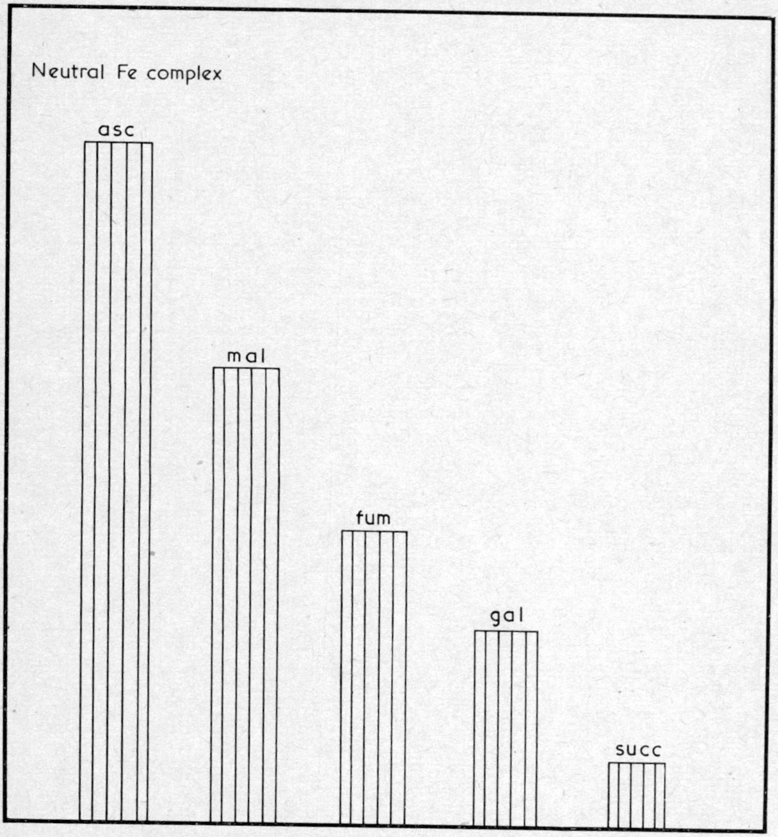

FIG. 43. The concentration of neutral complex present at pH 6–7 for the interaction of Fe(II) with ascorbate, malate, fumarate, galacturonate and succinate.

It is often necessary to insert metals surgically as structural components, for example, skull plates or hip pins and, of course, it could be disastrous if these were slowly to dissolve into body fluids. Thus, for these purposes, metals which are very 'soft' in the HSAB sense are chosen—for example, gold, silver, tantalum or platinum which, for reasons discussed in Chapter 4, are not water soluble.

Copper and rheumatoid arthritis

We have considered how ligands, complexes and metals may be absorbed, now let us consider a current problem in metal therapy. As we shall see, the situation may be much more complicated than our simple examples might suggest.

There is now a great deal of evidence connecting copper with rheumatoid arthritis and other inflammatory diseases but its actual function at the molecular level is not known.

Most of us will have heard the 'old wives' tales' that wearing a copper bracelet helps alleviate the symptoms of rheumatism, but only now are scientists giving credence to this idea. Studies are being carried out in Australia to determine how much copper can be absorbed, as salts of the components of sweat, through the skin from these bangles. A number of other folk remedies for arthritis happen to involve substances which are high in copper content, for example, shellfish, nuts, mushrooms or cider vinegar.

Rheumatoid arthritis is characterised by a high level of copper in the serum but most of this is bound to the serum albumin and ceruloplasmin and so is not pharmacologically active. It is believed that the drug D-penicillamine and perhaps other anti-inflammatory drugs, are able to mobilise this copper store in a manner similar to that explained in connection with Wilson's therapy (p 55). Some of this enhanced pool of low molecular weight copper will be in a form that can reach the inflamed tissue (this requires membrane penetration and so neutral complexes are desirable) and therein, through a mechanism that is as yet far from clear, it promotes a reduction in the inflammation. Clearly, it is not absolutely necessary to supplement the low molecular weight form of copper complex at the expense of the high molecular weight protein bound copper but rather direct supplementation using low molecular weight species is feasible. Examples of the benefits of this direct approach include the bangle copper perspiration complexes and also various low molecular weight forms of copper administered orally. These can sometimes reduce the intestinal ulcerative side effects of anti-arthritic drugs as well as improve their efficacy (*eg* copper aspirinate appears to be 20 times more effective than aspirin itself).

Thus pharmacologists not only have to find a complex *via* which copper can be absorbed but also have to decide whether this should be in the form of Cu(I) or Cu(II) and how the interactions with tissue and with human serum albumin will be affected.

In spite of the difficulties discussed above many metals are indeed used in therapy; a list of the more common ones is given in Table 19. As our knowledge of the *in vivo* function of trace metals increases this list will no doubt grow.

Concentration effects and dose-response relationships

As stated previously, the concentrations of metal ions are controlled (by proteins, buffer systems and hormones) in various organs and disorders arise when this fine balance is upset. The imbalance, and hence our method of treating the condition, may either be caused

Table 19. Some of the metal complexes administered as drugs in the UK.

Antacids
Sodium bicarbonate
  -silicate
Magnesium carbonate
  -hydroxide
  -trisilicate
Calcium carbonate
Aluminium hydroxide
Bismuth carbonate
  -aluminate
  -subnitrate

Gastro-intestinal sedatives
Sodium bicarbonate
  -alginate
Magnesium carbonate
  -oxide
  -trisilicate
Calcium carbonate
Aluminium hydroxide
  -sodium silicate
Bismuth carbonate
  -subnitrate
  -tripotassium di-citrato

Drugs acting on the rectum
Calcium chloride
Aluminium acetate
Bismuth oxide
  -subnitrate
  -subgallate
Zinc oxide
Titanium dioxide

Expectorants, cough suppressants, mucolytics and decongestants
Sodium percarbonate
Potassium guaiacolsulphonate
Calcium iodide
Copper sulphate

Local reactants on the nasopharynx
Silver—mild protein

Oropharyngeal preparations
Lithium acetarsol
Zinc bacitracin

Anti-inflammatory and anti-allergic preparations
Zinc sulphate

Skin soothing and protective preparations
Aluminium dihydroxy allantoinate
Zinc oxide
  -salicylate
  -10-enyl undecenoic salt

Keratolytics and cleansers
Selenium sulphide

Topical and anti-fungal and anti-infestive preparations
Zinc undecenoate
  -naphthenate

Topical anti-infective preparations
Bismuth subgallate
  -formic iodide
Zinc oxide

Drug dependence
Magnesium carbonate
Calcium tribasic phosphate
  -citrated carbimide

by a metal ion excess or deficiency in the diet, or by an imbalance in the organic biochemicals (*ie* ligands) controlling the metal ion balance. Thus, the atypical metal concentration may either be the cause of the disease or just one of the symptoms. In each case, the analysis of the body fluids is a good diagnostic indicator of the presence of disturbances. We shall amplify these observations by describing some examples of the regional incidence of abnormalities caused by *total* elemental deficiencies or excesses.

One example of a regional deficiency causing disease has already

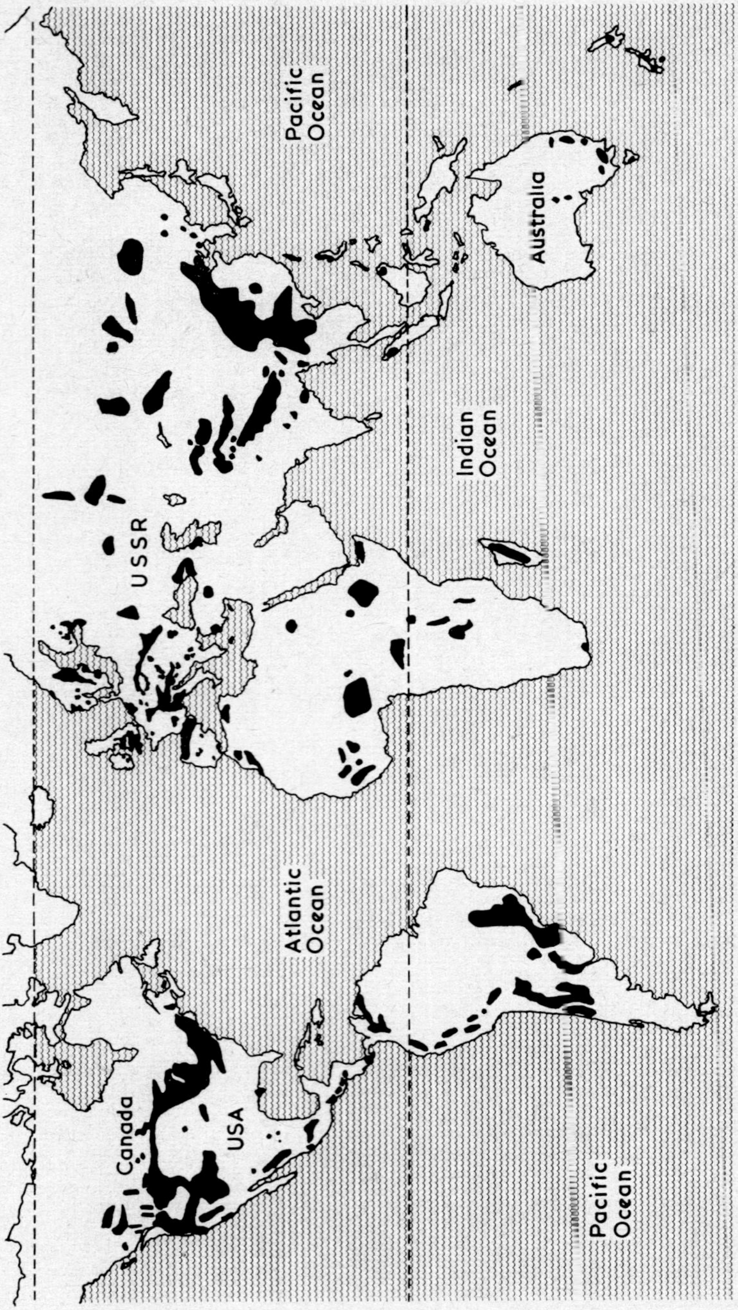

FIG. 44. Map showing the goitre areas of the world which are also areas of iodine deficiency.

been discussed in some detail in Chapter 5. There it has been shown that a soil deficiency of magnesium in some areas of Poland causes an increase in *Aspergillus flavus* which in turn increases the incidence of leukaemias. A better known example of a deficiency disease perhaps is the occurrence of goitre—the enlargement of the thyroid gland caused by a lack of iodine. Goitre affects both animals and man in the low iodine content regions shown in *Fig. 44*, however, it is one of the simplest deficiencies to combat, the addition of iodine to the table salt usually being sufficient for prophylaxis.

Regional excesses which lead to ill-health are becoming more common as the world becomes more industrialised, one important example being the 'itai-itai byo' (ouch-ouch disease) caused by chronic cadmium poisoning. Cadmium is in the same periodic group as zinc and so is present to some extent in all zinc ores. Itai-itai disease was found to be endemic in an area of the Toyama Prefecture of Japan (*Fig. 45*). The symptoms of this extremely painful disease are bone deformation and pain caused by the slightest pressure on the bones. In fact, multiple fracture of the ribs can be caused even by the pressure of coughing. Many careful studies had

FIG. 45. The Toyama Prefecture of Japan showing the area in which 'itai-itai' disease is endemic.

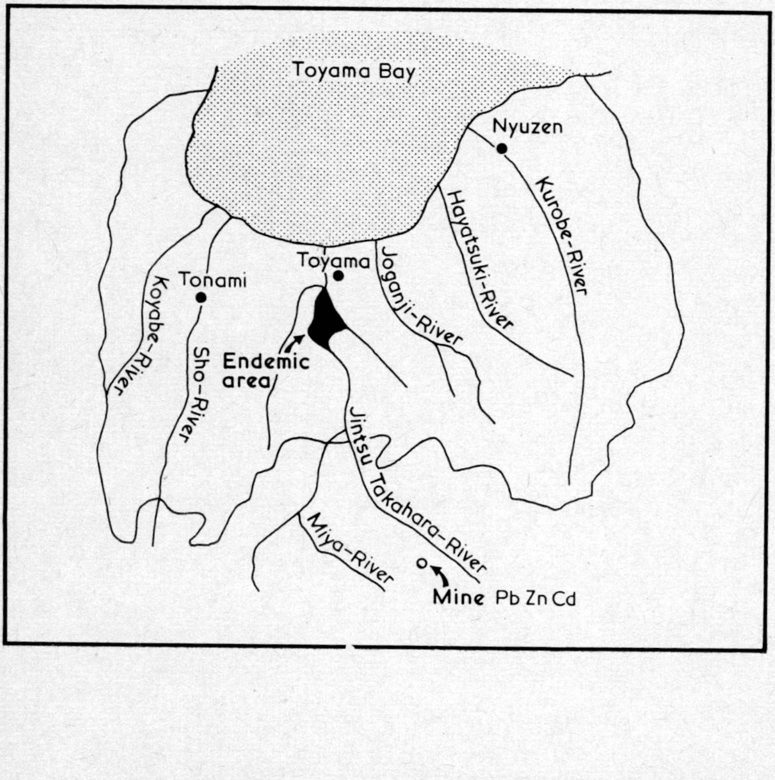

to be carried out to determine the cause of this disease but eventually cadmium was implicated and its source found to be the Kamioka mine. Faulty treatment of waste water from the mine polluted the Jintsu River with particles of a very high cadmium content. When this water was then used to irrigate the rice fields these particles settled out and the top soil built up an extremely high cadmium concentration. Unfortunately, rice can take up considerable quantities of this metal from the soil and the local people had a diet which was low in both calcium and vitamin D, all of this leading to a high incidence of 'itai-itai byo'.

The result of a toxic substance, be it a metal or a drug, will, obviously, depend on its concentration, but for any given dose there is also considerable individual variation. Thus we observe a Gaussian distribution as shown in *Fig. 39.* We also expect the response to increase as the dose is increased as in *Fig. 46* but this is by no means always the case. For example, for the bactericide oxine, discussed in Chapter 5, the response decreases as the drug dose is increased since it is the 1:1 complex with *in vivo* iron which is pharmacologically active, the 2:1 and 3:1 complexes being inactive.

FIG. 46. A normal drug dose–response relationship.

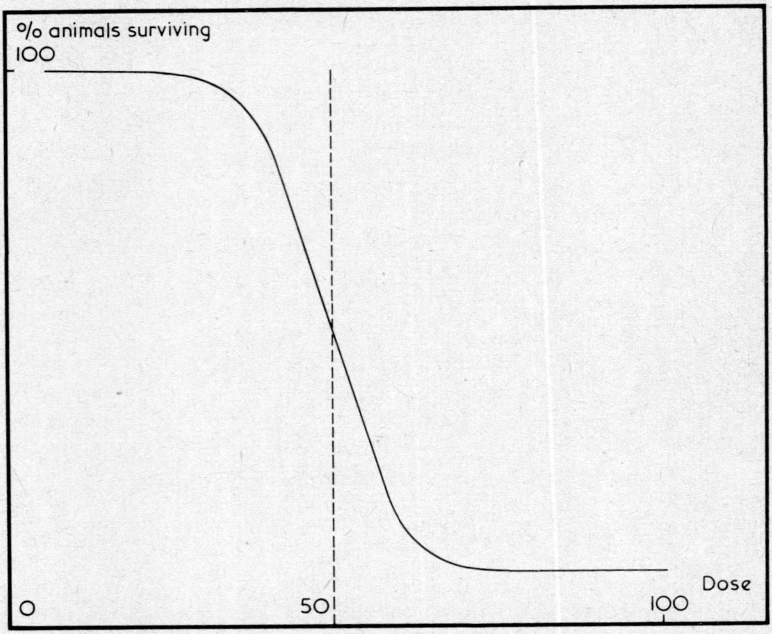

Mechanisms of drug activity

This section describes some mechanisms involved *in vivo* and indicates how a knowledge of the bio-inorganic chemistry involved can be used to influence and sometimes improve the processes *in vivo*. In general terms, we can discuss bioavailability from drugs and from food, the beneficial and toxic effects of drugs and the excretion of elements.

Most elements enter our bodies *via* plants (or animals which have eaten these plants) that have competitively complexed these elements from the soil. Soil itself may contain some fairly powerful chelating agents, for example, humic acids which are polyanionic substances containing many –OH and –COOH functional groups. The chelating agents from the plants have to compete with this complexing in order to allow the plant to take in the essential metals. The root hairs of some plants can, in fact, secrete chelating agents which solubilise compounds such as ferric oxide and calcium carbonate and so make the iron and calcium available for absorption.

In every country there are regions in which plants show the symptoms of deficiency of one or more of the trace elements. For example a lack of iron can cause the iron chlorosis or bleaching of the leaves shown on the citrus plant in *Fig. 47*. This may be caused

FIG. 47. A lack of soluble iron in the soil causes the bleaching of these citrus plant leaves (H. F. Walton, *Scient. Am.*, 1953, **74,** by kind permission).

by a deficiency of iron in the soil, in which case it can be supplemented by the addition of a soluble iron chelate, preferably a mixed ligand complex since this will give additional stability. On the other hand, there may be plenty of iron in the soil but in an insoluble form which is unavailable to the plant. This may be due to low levels of humus or potassium or to high levels of phosphate or hydroxide. In this case it is necessary to add only a chelating agent which will solubilise the metal, EDTA often being used in these circumstances.

Once man has eaten the plants, he must be able to absorb the essential elements through the gastrointestinal tract, but this is not always possible. The soya bean, for instance, is able to grow on very poor soil because it possesses powerful chelators with which to leach nutrition from this soil. However, when man eventually eats soya bean protein these chelators still bind the essential elements so strongly that they cannot be removed to allow absorption through the intestinal wall. This means that as the proportion of man's protein obtained from soya bean increases serious thought will have to be given to dietary trace element supplementation.

Each metal ion and also each administered drug is involved in a variety of mechanisms *in vivo*. We shall describe only three examples of diseases which can be treated using bio-inorganic drugs.

Cancer

Cancer is a disease characterised by the uncontrolled multiplication and spread within the organism of apparently abnormal forms of the organism's own cells. Our bodies produce about 10^{11} cells per day (about 500 g of new tissue), between 10^4 and 10^6 of which are imperfect, but these are generally destroyed by our immune defence system. One type of cancer or another is responsible for one in five deaths and leukaemias and lung cancers are becoming more common all the time.

The cause of cancer is by no means simple (*Fig. 48*). The steps in this process have not been clarified but carcinogens may be metals,

FIG. 48. The production of a secondary cancer.

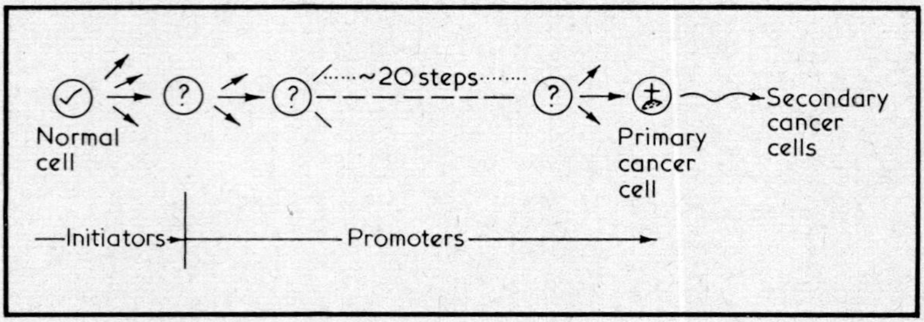

ligands or complexes and may get into the body by a variety of routes, *eg* the lungs, intestine or skin. The greatest challenges to health are from species having similar chemical properties to biologically essential species.

Many metals have been shown to be carcinogenic in test animals and it is known that the following metals all accumulate in our bodies during ageing: Al, As, Ba, Be, Cd, Cr, Au, Ni, Pb, Se, Si, Ag, Sr, Sn, Ti, and V. Indeed, as discussed in the previous chapter the distribution of some cancers may be correlated with top soil trace metal content. However, most cancers are caused by ligands, *see Fig. 49*, and a few are caused by viruses.

It is clear that the best long term approach to cancer is the elimination of environmental carcinogens and the frequent screening of susceptible individuals. One problem with cancer therapy is that

FIG. 49. Some known and suspected carcinogens.

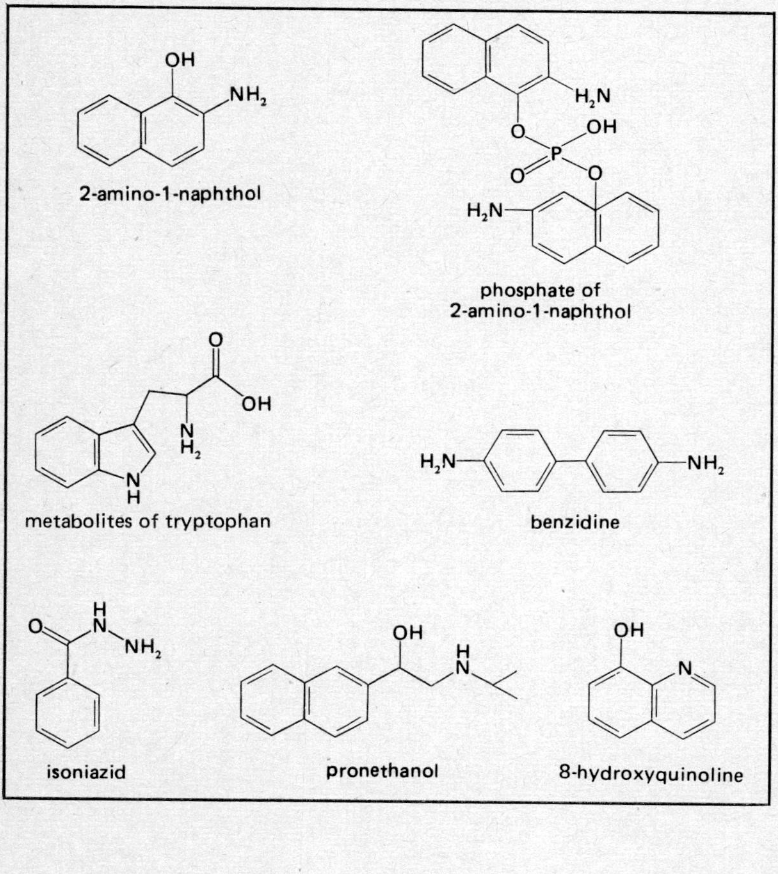

the presence of the disease is not realised immediately, the main reason for this being that cancer cells themselves contain no nerves and so no pain is felt until they start to put pressure on normal tissue. Often the stage at which a cancer is detected determines the therapy to be employed, *eg* surgery and radiation for localised tumours or chemotherapy for disseminated cancers.

The cancer clinician has relatively few drugs available for the treatment of inoperable cancer and none of these is specific for cancer cells but they are really antigrowth agents which also attack normal cells and so almost always give serious toxicity problems to the patient. The anticancer therapeuticals at present in use can be divided into four categories.

(*a*). Alkylating agents.

(*b*). Antimetabolites—these are similar to a chemical essential for growth but once a tumour cell has incorporated one of these replacements it finds that it cannot perform the functions necessary for further growth.

(*c*). Enzymes—some tumour cells for example are dependent on asparagine but cannot synthesise their own supply. Thus, if we can remove blood asparagine by the use of asparaginase we can effectively starve the cancer cell.

(*d*). Hormones—the administration of an opposite hormone can be used to treat tumours in organs under hormonal control.

It has been found that some complexes of Pt(II) or Pt(IV) are very potent anti-tumour agents, being effective against transplanted, carcinogen initiated and virally induced cancers. The complex diaminodichloroplatinum(II) caused complete tumour regression in mice even when it was withheld until the tumour was at an advanced state. Furthermore, the treatment also conferred immunity to rechallenge with the same tumour for up to 11 months. Anti-tumour activity is also found for the complexes shown in *Fig. 50*.

As can be seen, all of these complexes are neutral and so are able to cross membranes *in vivo*. Also, they all contain at least two adjacent reactive ligands such as chloride, the *trans* isomers of the same composition being completely inactive. It has been shown that their anti-cancer activity is due to the inhibition of the cancer cells' DNA synthesis, and the mode of action is thought to be the release of the chloride ions so allowing the platinum to form a cross link between the nitrogens of two purines on the DNA chain. Alkylating agents such as the nitrogen mustards, on the other hand, are thought to form similar crosslinks but, in this case, between bases on different DNA chains. The distances between the chloride ions on the platinum complexes, 0.33 nm, and on the nitrogen mustards, 0.80 nm, suit perfectly for the formation of intra- and interstrand bridges respectively.

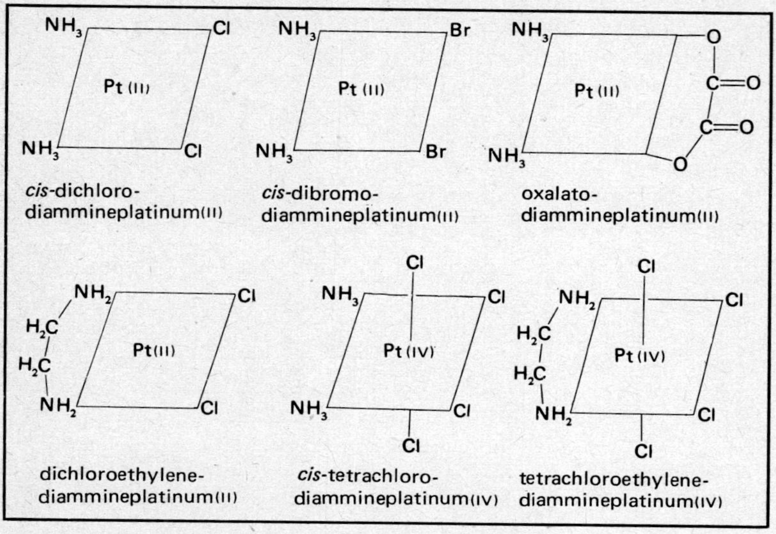

FIG. 50. Structural formulae of some active anti-tumour complexes of platinum.

Some cancers are able to repair these anomalous bonds individually but not when both types are present together, and so drug immunity can be overcome by administering a therapeutical containing both types of agent.

Viral diseases

As discussed above, viruses may cause cancers but they also cause a wide variety of other diseases the treatments for which are far from satisfactory.

A virus particle in isolation is unable to reproduce so it has to gain access to a host cell in order to be able to direct its own replication. The viral nucleic acid enters the nucleus of the host cell and forces it to stop the replication of its own components and move over to the synthesis of the components for the production of more viruses. It is believed that the attack of a virus on a cell is mediated through a metal ion. The cell wall of the bacterium *E. coli*, for example, contains a zinc complex to which a bond may be formed by a sulphur bridge of the virus's protein coating. This causes the splitting of the coating and allows the viral DNA to enter the host cell (*Fig. 51a*).

Viral replication may be prevented by removing the zinc from the bacterial cell wall with a strong chelating agent such as EDTA or by causing the viral coating to split and release its DNA before it reaches the host cell. This may be achieved by presenting the virus with an alternative metal complex to which it may bind—$Cd(CN)_3^-$ has been used for this purpose (*Fig. 51b*).

It has been postulated that viruses require optimum concentrations of copper and zinc in order to proliferate, these metals being essential for a large number of enzyme reactions. So, if we could drastically increase or decrease these concentrations, viral growth might be limited so offering the host's immune response system a favourable chance of overcoming the invasion. Indeed, if the optimum concentrations did not vary greatly between species of virus, then positive or negative metal therapy might well give a relatively wide-ranging treatment for viral diseases.

However, metal concentrations discouraging viral growth might also be damaging to the normal host tissue and many studies will be necessary in order to choose the supplementing metal complex, or the depleting ligand, which shows the maximum difference between the concentration toxic to the virus and the concentration toxic to the host.

FIG. 51. (*a*) The attack of a virus on a bacterium.
(*b*) The attack of a metal complex on a virus.

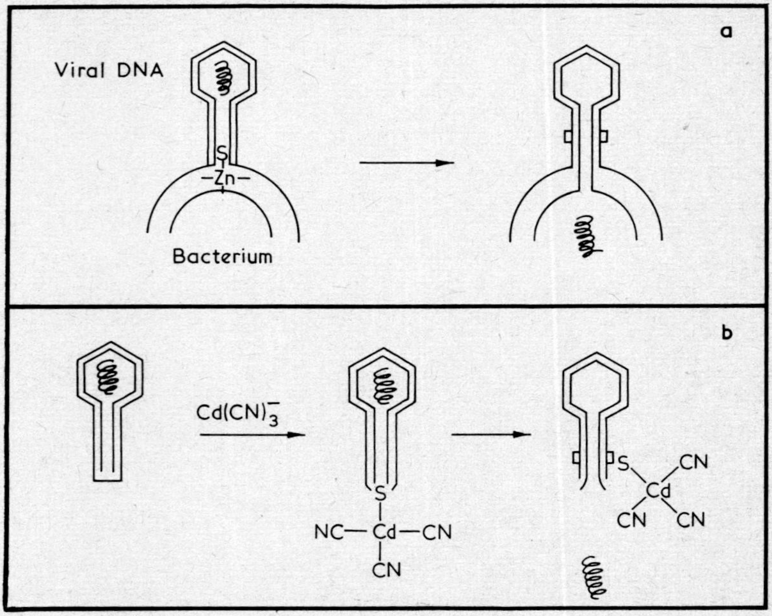

Ulcers

Many drugs containing bismuth are used in the treatment of disorders of the alimentary system, a colloidal bismuth citrate solution held at pH 10 by the addition of ammonia being frequently used in ulcer therapy.

The introduction of fibre optics has recently made possible intragastric colour photography and so the progress of the ulcer during treatment can be observed. These photographs show that the ulcer site becomes coated with a precipitate which isolates the underlying raw surface from the digestive action of the gastric or duodenal juices and allows healing to progress.

By making a series of *in vitro* measurements of formation constants and solubility constants and by devising computer simulation models of the equilibria present in the stomach it is possible to suggest the following mode of action: on the acidification of the bismuth citrate solution by the hydrochloric acid present in the stomach two solids are precipitated; bismuth citrate and bismuth oxychloride (*see Fig. 52*). There is an optimum pH for maximum precipitation and, in general, the stomach contents are below this pH. However, if amino acid anions and other bases are present in the fluid oozing from the ulcer site their donor groups will be protonated thus causing a localised lowering of hydrogen ion concentration and a raising of the pH towards the desired optimum. By this means the precipitation of the protective coating is localised around the ulcer site.

The computer models can be used further to predict the best bismuth to citrate ratio in the drug and the best dose level, or even to suggest a different formulation for the treatment of ulcers at other than gastric pHs, for example mouth ulcers which are bathed in saliva at a pH of about 7.4.

In the treatment of disease using metallotherapy we must keep in mind the fact that once the introduced metal has performed its useful function then it has to be removed from the body. If this is an essential/beneficial metal then a route of excretion will already exist but if a non-biological metal is introduced its removal may not be a simple process.

From the above examples we can see that the application of the principles of bio-inorganic chemistry can yield a great deal in terms of the understanding of disease. Better understanding of disease will, of course, always, eventually, lead to its better treatment.

Future developments and trends in bio-inorganic therapy

There are two facets to bio-inorganic health—the prevention of the imbalance and the treatment of the diseased person. We shall consider both these areas and then discuss other influences affecting the rate at which a drug comes to market. However, we ought to note

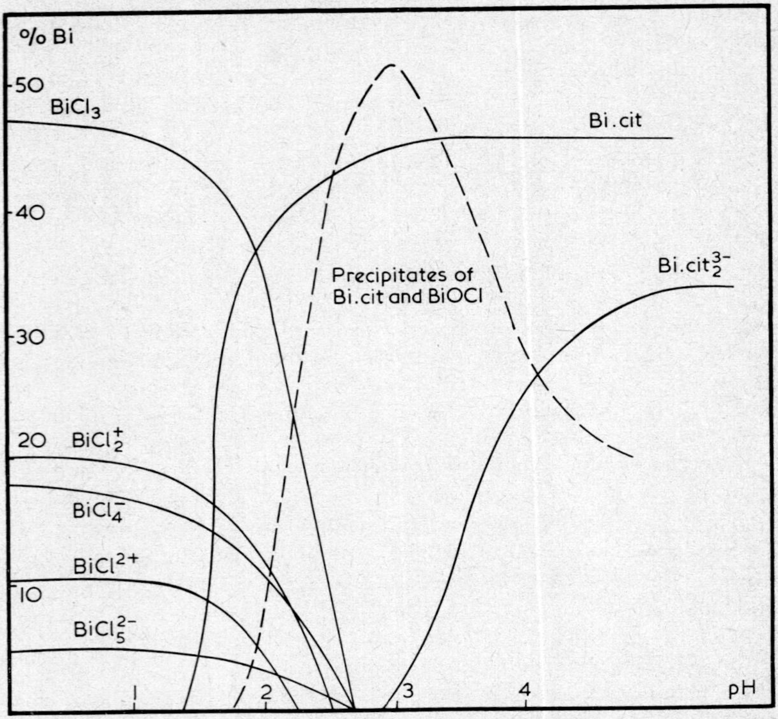

FIG. 52. A computed pH profile for the bismuth–citrate–chloride system for a single dose of anti-ulcer medicine diluted in the stomach fluid. The percentage axis represents the percentage of total bismuth in each form.

that we are discussing bio-inorganic aspects rather than the whole field of therapeuticals.

Prophylactic measures. The lessons to be learned from correlating the incidence of cancer with the concentrations of bio-available selenium (p 75) or magnesium (p 77) in the environment have been registered and an international cooperative enquiry was set up some two years ago to search for similar trends in diseases dependent upon the concentrations of cadmium, copper, chromium, magnesium, molybdenum and zinc. A secondary study will involve the eipdemiology in relation to the ratios (*eg* Mg:K or Cu:Zn) of these elements. Once a pattern of anomalous disease has been discovered judicious agricultural supplementation or sequestration can produce the required balances (equivalent to the fluoridation of water supplies in low fluoride areas and the addition of calcium salts to reservoirs in high fluoride areas in order to reduce the incidence of dental caries). In addition to controlling diseases caused by invading organisms (for example magnesium *versus* aflatoxins and fluoride *versus* streptococcus causing caries) this approach will broaden to consider other areas such as the chromium dependence of diabetes, the lithium, sodium, potassium, magnesium, calcium and lead dependence of psychotic disorders, and the magnesium:calcium ratio dependence of myocardial ischaemia (for example, the higher incidence of heart attacks in the soft water areas of north west England and in Scotland in comparison with the hard water areas of the south-east).

Yet another territory in which one hopes that preventative measures might be increased is in the field of radiation protection: of all the poisons entering the human body, some of the radioactive elements are undoubtedly the most dangerous, in particular isotopes of plutonium, strontium and caesium. Chelating agents related to EDTA, prussian blue and sodium alginate have been found useful in preventing their absorption from the stomach. These protective ligands are chosen such that they form insoluble complexes with the radioactive metal ions and so, whilst permitting them to pass through to the faeces, block off their passage into the blood and bone marrow. In principle, it ought to be possible to design agents more akin to endogenous ligands so that they could be used on a daily prophylactic basis by exposed personnel, *ie* as special dietary supplements for people at risk. This is an extension of the principle of heat-exposed workers taking additional sodium chloride.

In similar vein, chelating agents capable of chelating copper(I), for example sulphydryl groups, are known to protect animals against lethal doses of ionising radiation.* In general, radiation is oxidising whereas it is copper(I), the reduced form of copper in solution, which is required by metalloenzymes for cellular repair processes. Similarly, chelating agents related to EDTA have been used to prevent cardiac arrhythmia by reducing not the oxidation state but the concentration of calcium ions in blood plasma.

Turning our attention to treatment, one must remember that the process of finding drugs for treating specific diseases is still largely fortuitous. However, therapeutic innovation in the next decade must surely be expected to make more use of *quantitative* measures such as the computer simulation of biological systems examples described earlier and such as chronotherapy. Our suggestion of the expansion of simulation is based upon two mutually augmenting reasons: (*a*) medical research is currently undergoing a transition

---

* A hypothesis currently in vogue is that the prophylaxis and treatment of the common cold using vitamin C (ascorbic acid) functions through the ascorbate reducing cellular Cu(II) to Cu(I) and thus permitting the damaged cells to be repaired. The redoxing system involving oxidised and reduced glutathione is thought to be an intermediary in this process.

from the cellular to the molecular level and so it is becoming feasible for chemists to design compounds with prescribed biological profiles in many more diseases than was possible in the past. (*b*) Secondly, computers can be used for preliminary screening and for formulations not easily established using animal models (for example, optimising the thickness of bismuth coatings in ulcer therapy).

In the not too distant future one can anticipate the availability of computerised data banks containing models of, say, plasma, cerebrospinal fluid, *etc*, so that the serum analysis for a polluted individual can be used as input, the model run and a reliable dose calculated for both the sequestering agent specific for the pollutant and for the names and amounts of essential/beneficial metals to be brought to balance by 'topping-up' therapy.

There are many other factors which influence how and whether a drug eventually comes to market. Finance plays a major part and we cannot escape the fact that the maximum profit accrues from maximum sales. For example, UK sales within the last decade have had the general order

> antibiotics > tranquillisers and anti-depressants > corticosteroids > cardiovasculars > anti-arthritics > cold preparations > analgesics = diuretics = sedatives = oral contraceptives > antacids = vitamins > anti-anaemics = antibacterials = dermatology preparations.

Clearly, this pattern reflects the incidence of disease and so no amount of keen marketing activity and superb drug design, even of a 'wonder drug' with negligible side effects, can change this order. Thus, new anti-arthritic agents are far more likely to be marketed than are Wilson's disease ligands (one or two cases per million of the population) even though both involve the same principle of the pharmaceutical manipulation of the body's exchangeable copper ion concentrations.

Drug discovery and introduction is a very laborious and, therefore, a very expensive (about £5m per new drug) performance. Peculiarly, the research is not necessarily the most expensive stage; screening, development (*ie* toxicity, side effect studies and scaling up to full production), and the submission of applications to the Government Committee on the Safety of Drugs can all be very costly.

For the future, one must differentiate between developing better *foods* or better *drugs* (the borderline is hazy) which are synonymous with the *prevention* and *treatment* of disease, the former terms implying correctly balanced diets in terms of trace elements and vitamins and even the intake of new vaccines, antitoxins and antisera (in principle these are not bio-inorganic but in order to boost effectively the immune response system the correct inorganic nutrition is necessary). Most likely there will be a greater emphasis on researching the basic biochemical aspects of cancer, kidney malfunction, pollution and industrial hazards, alcoholism, birth control, mental and nervous disorders and heart and circulatory disease and it is hoped that this will lead to new methods of treatment. Quantitative drug design and computer simulation must, of necessity, be used in ever increasing amounts to reduce innovation costs by assisting in the search for 'lead' molecules and by rationalising animal screens.

Many of our remarks are generally applicable to the whole drug research scene but we confidently predict that, because the roles of bio-inorganic balances have been neglected for so long, the impact made by this subject in the next few years will be iconoclastic. There is one final prediction that we wish to make: there are three properties of future researches that are already in evidence—they will surely gravitate towards being more expensive, more controversial and more keenly focused on the increasing market for drugs to treat the so-called civilisational or sociological conditions caused by pollution overloading the environment or our emotional behaviour overloading vital organs.

7. Prospects

Previous chapters have tried to demonstrate that bio-inorganic chemistry is an embryonic branch of natural science which is rapidly gaining momentum towards a fascinating and exciting future. The history of the subject may be traced back to the origins of life on Earth when the survival of primitive species demanded that they took advantage of all of the elements available on the Earth's surface and of their respective chemical properties.

Healthy life is still very much dependent upon this interaction between an organism and its environment (and also the element contents of other species below it in the food chain). Much of this environment is 'inorganic' (*ie* it involves elements in addition to carbon, hydrogen, oxygen and nitrogen) and so there is a growing appreciation of the roles of these inorganic elements in man.

Thus, it is not surprising that malfunctions in man's bio-inorganic chemistry can often be traced to the evolutionary origin. Clearly, we cannot retrospectively modify our history but we can, to a certain degree, control future developments and also influence the malfunctions at the molecular level. Brown's monograph lists and describes the medical conditions most in need of research and solution: immunological rejection processes, rheumatoid arthritis, viral and bacterial infections, cancer, ageing and cardiovascular disease. Agriculture and food production also have areas in need of improvement. All of these subjects can be profitably researched from a trace element point of view. Table 20 suggests some of the

Table 20. Important factors to be considered when first discussing a bio-inorganic phenomenon.

Evolutionary aspects?	Biphasic response?
Ionograms in man and nature?	Essential or contaminating?
Cycles in nature?	Oxidation states?
Bioavailability?	Lability/inertness of complexes?
Circadian rhythms?	Stimulations/antagonisms?

more important principles to be considered at the start of any discussion; they have all been described in earlier chapters.

Bio-inorganic chemistry is a very complex subject to approach from first principles and yet such complexity is inherent in all biological subjects and has not prevented the discovery of their basic principles. Perhaps we have tended to overemphasise the aqueous aspects of bio-inorganic chemistry whereas many metals are, of course, involved in essentially non-aqueous environments in the active sites of metalloenzymes. On the other hand, an over-fascination with metalloenzymes may detract from the broader perspective of the complete cycles of elements *in vivo*, *ie* that any given metal ion does not exist *ad infinitum* in a given site but rather it progressively participates in mechanisms in various parts of the animal. A complete study of an element's bio-inorganic chemistry must necessarily involve the integrated efforts of many specialists in addition to, say, metalloenzyme crystallographers or solution molecular biophysicists. Without taking such an overall view we shall continue to miss the forest for the trees on many important occasions.

Scientists have a particular involvement in the life and future of *homo sapiens*. Life, of course, has many facets and can be defined rather quantitatively in terms of its survival rate or perhaps more humanly in terms of its quality. What are the aims of our Society? —growth, short term prosperity, knowledge for knowledge's sake regardless of sociological consequences, efficiency, self glory, the joy of discovery, peace of mind—which takes priority and what role should science play? On the one hand, the vast majority of readers would question the statement that 'man's aim is to conquer nature in order to exploit her for his own ends' and yet, on the other hand these same tax-payers demand the maximum short term yield per pound invested in shaping nature to suit our health, educational, nutritional, transport and defence requirements. Truly, although bio-metallic manipulation of life processes to produce a better life is becoming feasible, it is first necessary to agree on a definition of 'better life'.

Meanwhile, we have seen that computer based models can be used to delete our ignorance of structure/activity and concentration/activity relationships and also contribute in the field of quantitative medication (dose optimisation). Another bio-inorganic involvement is that of the choice of trace elements to be added to the increasing volume of artificial foods. However, history can teach us a lesson here—at the beginning of this century the most useful tool possessed by biochemists was that cabbalistic scribble of letters and connecting lines with which they pictured molecules. Next, co-ordination chemistry and chelating agents became popular, these leading on to computer based pH profiles but we must note well that just as the exactness of chemical formulae and the specificity of chelating agents are remembered today, so too our efforts at cranking a computer will be remembered not only for the hypotheses that they produced but also for the precision of the input data. There is no substitute for accurate results.

The quantity of data will undoubtedly expand as additional roles are found for the known bio-elements and as the lists of essential, beneficial or polluting elements enlarge.

Although there is a great deal yet to be understood concerning the effects of trace elements in human physiology and many benefits to be obtained from this knowledge, much of the research will be sterile unless there is a narrowing of the gap between research conclusions and clinical applications and also an increased effort to keep the general public informed.

Bio-inorganic chemistry is too important a subject to be left to chemists, the concepts must be discussed with other scientists and they, in turn, must be prepared to discard parochial attitudes and to have interests in addition to their primary discipline.

Scientists are amongst the few fortunate people who can still find peace and contentment in work that is well done, that suits their interests and capabilities and that is directed at altruistic objectives. So, in the past, the inorganic chemist has probably benefited more from the attraction of bio-inorganic medicine than it has from him, but all this is now changing and bio-inorganic chemists can make worthwhile contributions to the maintenance of health and the treatment of disease.

Suggestions for Further Reading

Chapter 2

D. R. Williams (ed.), *An introduction to bio-inorganic chemistry*. Springfield, Illinois: Thomas, 1976; see *chapters* by D. R. Williams, R. Österberg, B. Sarkar, D. D. Perrin and R. P. Agarwal.

I. Zipkin, *Biological mineralization*. New York: Wiley, 1973.

E. Ochiai, *J. chem. Educ.*, 1974, **51**, 235.

C. L. Hamilton (ed.) *Chemistry in the environment*, Readings from *Scientific American*. San Francisco: Freeman, 1973.

Chapter 3

A. L. Lehninger, *Biochemistry—the molecular basis of cell structure and function*. New York: Worth, 1970.

D. R. Williams, *The metals of life*. London: Van Nostrand Reinhold, 1971.

I. J. T. Davies, *The clinical significance of the essential biological metals*. London: Heinemann medical, 1972.

J. M. Pratt, *Inorganic chemistry of vitamin* B_{12}. London: Academic, 1972.

M. M. Senozan, The Chemical Elements of Life, *Scient. Am.*, July 1972, **227** (1).

J. T. Spence, Biochemical Aspects of Molybdenum Coordination Chemistry, *Coord. Chem. Rev.*, 1969, **4**, 475–498.

Chapter 4

D. R. Williams (ed.), *An introduction to bio-inorganic chemistry* (*see* details above).

J. E. Huheey, *Inorganic chemistry: principles of structure and reactivity*. New York and London: Harper and Row, 1972.

M. N. Hughes, *The inorganic chemistry of biological processes*. London: Wiley, 1972.

H. Sigel (ed.), *Metal ions in biological systems*. New York: Dekker, 1974.

F. A. Cotton and G. Wilkinson, *Advanced inorganic chemistry*, 3rd edn. New York: Interscience, 1972.

P. M. May, P. W. Linder and D. R. Williams, Computer Simulations of Biological Fluids, *J. chem. Soc.*, *Dalton*, 1977, 188; *Experientia*, 1976, **32**, 1492; *FEBS Letts*, 1977, **78**, 134.

Chapter 5

E. G. Brown, *An introduction to biochemistry*, Monograph number 17. London: RIC, 1971.

E. Krőrös, Oscillating Reactions, and R. Österberg, The Origin and Specificity of Metal Ions in Biology, in *An introduction to bio-inorganic chemistry*, D. R. Williams (ed.) (*see* details above).

M. D. Walker and D. R. Williams, The Chemical Principles of Chronotherapy as Established from an *in vitro* Model of Circadian Concentration Rhythms, *Bio-inorg. Chem.*, 1975, **4**, 117.

A. Furst, *Chemistry of chelation in cancer* Springfield, Illinois: Thomas, 1963.

D. R. Williams, Metals, Ligands and Cancer, *Chem. Revs.*, 1972, **72**, 203.

D. R. Williams, Anticancer Drug Design Involving Complexes of Amino-acids and Metal-ions, *Inorg. Chem. Acta. Revs.*, 1972, **6**, 123.

Chapter 6

W. Breckon, *The drug makers*. London: Eyre Methuen, 1972.

J. N. T. Gilbert and L. K. Sharp, *Pharmaceuticals*. London: Butterworths, 1971.

A. Goldstein, L. Aronow and S. M. Kalman, *Principles of drug action—the basis of pharmacology*. New York: Harper and Row, 1968.

F. P. Dwyer and D. P. Mellor, *Chelating agents and metal chelates*. New York: Academic, 1964.

Glossary

Aerobic, in the presence of air or oxygen.
aetiology, the study of the causes of disease.
anaerobic, in the absence of air or oxygen.
ataxis, lack of coordination between muscular movements.
atherosclerosis, thickening and hardening of the walls of the arteries.

Ceruloplasmin, a copper protein containing eight atoms of copper per molecule, having a molecular weight from 130 000 to 160 000 Daltons and existing in blood plasma at 0.3 g l^{-1}.
circadian rhythm, daily concentration or behavioural cycle.
cofactor, the extra molecule or ion that is necessary before an enzyme can function.
collagen, a protein which is one of the components of the body's fibrous tissues.

Diathesis, a liability or tendency towards a particular disease.

Entatic, strained, *ie* chemical bonds displaced from their 'normal' configuration associated with a metal ion.
epithelial, skin and skin-like cells covering the surfaces of the body and internal cavities.
exudative, the passage of blood components through the walls of arteries and veins into adjacent tissues.

Fistula, an unnatural, narrow channel leading from some natural cavity, for example the interior of the intestine or bowels, to the surface.

Haemochromatosis, a disease characterised by cirrhosis of the liver and diabetes associated with pigmentation from deposits of iron compounds. It often originates from the excessive absorption of iron from the intestinal tract.
homeostasis, the process of maintaining constant physical and chemical conditions within the body in spite of external changes.

Ischaemia, bloodlessness of part of the body caused, for example, by a blocked artery.

Keratin, the protein which forms horny tissues, hair and nails.

Mitochrondrion, the intracellular body responsible for oxidation of carbohydrates, lipids and amino acids and the site of the electron transport chain.

Necrosis, death of a limited portion of tissue.

Oncogenic, tumour-forming.

Pathological, of the nature of, or arising from, disease.
polycythemia, an increase in the number of red cells in blood.
prophylactic, an agent which prevents disease.
prosthesis, artificial parts to the body *eg* a limb, an eye or dentures.
pyrexia, temperature rise, fever.

Siderosis, deposition of excess iron in the tissues.

symbiosis, the living together in close association of two organisms of different kinds for their mutual benefit.

Urticaria, nettle rash, an allergic reaction.

Wilson's disease, a progressive disease of the liver and nervous system associated with an imbalance in one's copper biochemistry.